Crime, Security and Surveillance

Crime, Security and Surveillance

Effects for the surveillant and the surveilled

Gudrun Vande Walle
Evelien Van den Herrewegen
Nils Zurawski

Eleven international publishing
The Hague
2012

Published, sold and distributed by Eleven International Publishing
P.O. Box 85576
2508 CG The Hague
The Netherlands
Tel.: +31 70 33 070 33
Fax: +31 70 33 070 30
e-mail: sales@budh.nl
www.elevenpub.com

Sold and distributed in USA
Lyceum Books, Inc.
5758 S Blackstone Ave
Chicago, IL 60637, USA
Tel.: + 1 773 643 1902
Fax: + 1 773 643 1903
E-mail: lyceum@lyceumbooks.com
www.lyceumbooks.com

Eleven International Publishing is an imprint of Boom uitgevers Den Haag.

Printed in The Netherlands

ISBN 978-94-90-94765-1
NUR 741

Table of contents

Introduction: Crime, Security and Surveillance – effects for the surveillant and the surveilled

*Evelien Van den Herrewegen, Gudrun Vande Walle and Nils Zurawski**

This book aims to provide a platform to discuss three subjects that currently hold a very prominent position, not only in criminology, but much further afield, namely crime, security, and especially surveillance. As the Canadian criminologist Kevin Haggerty noted during a conference in London in 2010, 'security is not the ownership of criminologists anymore'. The attendees of that particular conference of the Surveillance Studies Network were an affirmation of his statement. Over 150 people from disciplines such as sociology, anthropology, geography, information science and technology, law and philosophy came together to debate, analyse and discuss issues around surveillance, many focusing on crime and security.

In 2009 an effort was made to form a network for all these scholars to interact, debate and engage with one another beyond their own disciplines and on a continuous basis. The COST action,[1] 'Living in Surveillance Societies' was formed and has since established itself as one of the significant forums concerned with academic engagement on surveillance within Europe. This edited volume, too, is a product of those collaborative efforts that make the network – through constant meetings, exchange of ideas and effective cooperation. The work of the network is structured around four working groups, one of which is concerned with examining citizens' everyday experiences of, and attitudes towards surveillance, out of which this edited volume emerged and took shape. It can be viewed as a result

---

* Evelien Van den Herrewegen is currently post-doctoral researcher at University of Antwerp, lecturer at Courtrai University College, and member of the Research Unit Social Analysis of Security (Ghent University).

Gudrun Vande Walle is assistant-professor at University College Ghent, faculty of Business Administration and Public Administration, Research Unit Governing and Policing Security; and Visiting Professor at Rühr University of Bochum, Master in Criminal Justice, Governance and Police Science. gudrun.vandewalle@hogent.be.

Nils Zurawski is professor and senior researcher at the Institute for Criminological Research in Hamburg. He is a member of the LiSS Management Committee and Coordinator of LiSS Working Group1.

1. COST or European Cooperation in Science and Technology is a European instrument supporting cooperation among scientists and researchers across Europe. For more information about the COST action Living in Surveillance Societies, we refer you to: http://www.liss-cost.eu/

of many of the working group's discussions and the common interests of its members on how surveillance impacts the everyday lives of people, bringing together often quite different discourses and understandings of what surveillance means, how it should be analysed or what perspective can be taken for this analysis. Also, the members of the network find themselves confronted with a variety of science cultures that add to the complexity of the interaction. Hence, this volume is also the result of bringing together different viewpoints from various disciplines and from around Europe.

Despite the fact that the term surveillance is more of a generalisation than a clear definition of the subject matter, the past decade has yielded a vast body of literature discussing it in depth and from quite different perspectives. As Haggerty & Ericson noted (2006, p. 21), surveillance as an analytic category is a simplifying device that suppresses a world of specific differences under a broad rubric. Even if the personal gaze (or that through a camera) was ever a common feature of analysis on surveillance, it has given way to other techniques and technologies such as sensors, statistics, satellites for such different purposes as climate watch, migration control, observation of body functions, management of prison populations, risk analysis and command and governmental techniques. Despite all these differences and ambiguities the term holds, it seems that one feature is shared by almost all phenomena that can be included under this subject or are connected to the issue: surveillance establishes a relation, i.e. it forms a division into an object and a subject, between something or someone being surveilled and something or someone surveilling. This dichotomic relation seems to be part of almost all forms of surveillance, with the possible exception of self-surveillance, e.g. during diets or in the contexts of sport, training and so on. However without going into a deeper philosophical discussion on self-referential modes of surveillance, we can also assume that such modes may also be constituted by or reliant on third parties or external agents, hence forming a binary relation of object and subject. Such a relation, established in whatever context and for whatever reason or aim, is far from being stable, fixed and distinct. Thus it seemed appropriate to dedicate this volume to that relation and its implications, especially in the context of everyday life experiences.

While for analytical purposes such a dichotomy seems quite useful, a closer look will show that it is more complicated than that. Surveillance is never only one-way, the relation is bi-directional at least, if not multifaceted. Surveillance in this relation becomes an enabler of communication or any other form of social interaction, not necessarily on equal grounds. It is a mediator, but also a field that structures its participants. To ask for its impact, on both the surveillant and the surveilled, implies that we need to talk about practices of surveillance, to view surveillance as an action that takes place, that is shaped by using it, that may be altered and one that is attached with meanings, which are neither given, nor self-evident.

The following contributions set out to reflect on the dichotomy of surveillance in the domain of crime control and security, and on the consequences of technologies employed in the course of a particular surveillance or security practice. Because of the complex nature of the relations established by surveillance practices and measures, any form of reflection needs to go beyond the simple questions of whether something might work, or if this or that technology has a panoptic effect, like so many popular accounts on the subject continue to ask. If we want to take the assumed dichotomic relation of surveillant and surveilled as a starting point and as the common core of all forms and modes of surveillance, then we need to know more about the social relations between different actors in surveillance settings, practices or regimes. We need to study the narratives, the work routines, the politics around surveillance, its technologies or any wish and hope that may be projected upon a camera, a biometric control device or a new policy. One way of doing this is to investigate the role of surveillance in the everyday lives of people – to ask about their concerns, their experiences and the contexts in question, i.e. in our cases contexts of crime prevention, criminal justice and the wider field of security, in which surveillance is becoming a more and more important aspect. So to ask about effects does not entail speculating about legal or philosophical impacts, but to look at actual experiences concerned with the politics and consequences of a technology, legal regulation or work routine.

Although each contribution deals with a specific aspect of surveillance, we have distinguished four broad themes and structured the book accordingly : 1. the responsibilisation of non-police actors in the domain of security; 2. nodal governance and nodal orientation of policing security; 3. policing public space and the consequences for the surveillant and the surveilled; and 4. the perception of surveillance. In what follows we give a short introduction and overview of the themes.

1 Responsibilisation

An important feature of the surveillance domain is that state agencies, i.e. the police, no longer position themselves as the main provider of security, but are actively and openly seeking the role of new types of actors to address crime and anti-social disorder in the public and quasi-public domain. As such, a variety of public (e.g. local administrations, schools) and private agencies (e.g. insurance companies, shopkeepers) but also private citizens are asked to take up the responsibility to monitor and scan their environment for potential risk and threatening behaviour, and take the necessary precautions to protect themselves, their customers and the public at large from harm. A number of contributions in this book have focused on this topic and examined how these 'new' actors deal with this appeal for responsibilisation.

In the first two contributions, the research is set in the Dutch supermarket. In the contribution by Patrick Van Calster and Marc Schuilenburg, the focus is on the 'Collective Shop Ban', a practice in which shopkeepers are given the opportunity to ban individuals form their shops if they exhibit undesirable conducts. This 'Collective Shop Ban' is an effort of the state to make shopkeepers co-responsible for maintaining security and preventing anti-social behaviour. In their research, the authors examine how the ban is applied in daily practice and how shopkeepers deal with this new responsibility. Although the Dutch policy-makers claim that the Collective Shop Ban is a success, the results of this research are somewhat disturbing as they clearly show that the ban is applied in numerous gradations and with an ample dose of arbitrariness, i.e. with shopkeepers who apply the measure very strictly, and others with more leniency. This questions the practice of the 'Collective Ban Shop' as regards the constitutional rights of the private citizens. As such the research of Van Calster and Schuilenburg raises important concerns about the legitimacy of responsibilisation for tackling anti-social behaviour.

Francisca Grommé also set out her research in the Dutch supermarket, but focused on how new technology interacts with this state's agenda to responsibilise new actors in policing activities. In particular, the author has made an in-depth analysis of the way supermarket employees and/or security guards use the technology of facial recognition to detect known shoplifters. Her conclusions are very pertinent for surveillance studies as her study affirms that, to understand the effect of technology, we need to study how these technologies are integrated within the relations, strategies, techniques, norms and preferences at a specific site. Her study shows that facial recognition was far from a neutral technology as it was an effort by the state to further enlist the supermarket in monitoring petty crime. However, she also notices that the effort was unsuccessful as the users of the technology refuse, question or resist the state's logic, and even adapt the new technology to their own logics of surveillance. As such, Grommé's conclusion also puts into perspective the appeal of responsibilisation: 'Surveillance technologies can potentially mobilise new actors, but they do not necessarily do so as new actors may refuse to take part in a particular mode of surveillance.'

Likewise, Evelien Van den Herrewegen raises some questions about the effects of this responsibilisation, thereby focusing on the perceptions and actions of 'normal' citizens. Based on empirical research, the author examines the assumption that nowadays people are caught up in a heightened crime awareness and more susceptible to the appeal of responsibilisation. In particular, she questions the assumption that this engrossment with security and individual responsibility has a detrimental effect on people's daily lives, and manifest itself as a problematic perception of safety. The research shows that, although people are indeed more crime conscious and feel responsible for their own safety and that of others, they do not automatically experience this as an assault on their everyday life. Moreover, people seek out strategies to predict, to avoid and even to control criminal victimisation, but also deploy some 'untraditional' strategies, such as fatalistic attitudes, to prevent this preoccupation with 'being safe' and its appeal for responsibilisation from having a detrimental effect on their daily lives.

The 'normal citizen' is also the focus in Agneta Mallén's contribution that examines the consequence of citizen journalism, i.e. citizens who use their mobile cameras to record and disseminate criminal acts. In this sense, citizen journalism can be viewed as a particular form of responsibilisation in which control is shifted from public organisations and agencies to individuals in their everyday lives. Mallén looks at the consequences of this citizen journalism both for the object under surveillance – in the shape of integrity risks and disintegrative shaming – and for the citizen journalists themselves – panopticism, crime prevention, and crime detection. In the analysis, the author makes an interesting comparison with the forms of control exercised by means of surveillance cameras. The analyses shows that there are indeed panoptic features to citizen journalism as surveillants can exert control at any point without the surveilled being aware of it. Turning to the surveilled, the author finds that citizen journalism raises concerns about privacy as it dissolves the boundaries between private and public, and that the citizen journalist control facilitates labelling that can lead to disintegrative shaming.

2 Nodal orientation and the nodal-network fallacy

Next contributions discuss the changed strategies of policing in a complex society of global mobility and the virtual character of flows. Nodal orientation is a concept based on the Network Society of Castells in which he elaborates the nodal outlook and has further been elaborated by Wood and Shearing (2007) in their nodal governance framework. Nodal orientation stresses the central function of the police that sees themselves confronted with other complexities. In nodal governance the police as a node have to collaborate with other nodes of non-state actors.

Marleen Easton and Fien Gilleir focus on those who carry out surveillance of transit migrants who enter Belgium via the seaports of Oostende and Zeebrugge. The logic they employ, the tasks to which this gives rise, and their collaboration is explained on the basis of qualitative empirical research in which concepts such as nodes and networks are used. Consideration of the 'nodal-network fallacy' makes it clear that various rationalities can stand in the way of an effective collaboration between the numerous nodes and actors present. Analysis shows that, on the one hand, there are many attempts for cooperation between several 'nodes/actors' but that, on the other hand, a great many obstacles, such as conflicting agendas and financing problems, stand in the way of the development of a network structure and thus also an effective approach to the challenges, also to security, posed by the arrival of transit migrants.

The Dutch authors Peter Marks and Arie van Sluis further elaborate the concept of nodal orientation as specifically interpreted by the Dutch Board of Police chiefs. While Easton and Gilleir consider nodal governance as a collaboration of plural security and non-security actors as elaborated by Clifford Shearing, nodal orientation refers to the police as a central player in security networks that control infrastructures where flows of people, goods and money move. This nodal orientation to control complex security flows changes the strategies of the police. It is a permanent search for the equilibrium between gathering intelligence and protecting privacy. The authors illustrate the changed functioning of the police in the Port of Rotterdam and at Schiphol airport.

Also Charlotte van Ooijen and Meike Bronkhorst study the way the police are dealing with nodal orientation. They specifically discuss the technological technique of Automatic Number Plate Recognition (ANPR) that has found its way into the Dutch practice of policing and crime investigation. One condition for nodal orientation is intelligence-led policing that has a large information database at its disposal. ANRP enables the police to register information that may be useful for criminal investigation. The authors explore to what extent the police themselves make an effort to pose the normative questions of legitimacy of ANPR policy and what kind of legitimacy takes priority. The text is the result of an empirical research conducted in one of the largest police forces of the Netherlands.

3 Policing public space

Another theme that has been discussed is the way in which securitisation has had an impact on the interaction of the surveillant and the public domain as its working space. Since the 1980s, public space has being used as a laboratory for securitisation: a diversity of private security actors and also citizens have joined the police in taking care of that public space, all from a different angle; new technology has been introduced for surveillance and detection while policing public space aims now more to avoid disorderly behaviour than to detect crime. All these changes resulted in an urban space that is for the one citizen a bit more public than for the other.

The authors mainly explore the changed expectations and attitudes of the police in this securitisation of public space. Antoinette Verhage and Paul Ponsaers wonder if this securitisation of public space can go together with the Community Oriented Policing philosophy which entails making policing more informal. They find that the police betray that COP philosophy and are taking a step back to more thorough public order policing.

Anaïk Purenne also discusses the role of the police based on a case study in two different forces of the Paris police department. Police officers in urban areas have become risk managers aiming to prevent the occurrence of danger. To enable risk management, the police are highly dependent on information gathering. The question Purenne raises is whether this risk-oriented policing has changed the police from an organisation using force as its main weapon into an intelligent police force that mainly gathers information. Another issue is whether information gathering legitimises the use of more force.

4 Reconsidering securitisation

Social scientists conducting research in the field of securitisation have mainly stressed its negative impact on society and on human rights. In the last three contributions the authors reconsider the consequences of securitisation, stepping away from the mandatory negative undertone. They try to grasp the advantages of securitisation and the borders of moral acceptability in three different domains.

Helena Machado, Susana Silva and Manuela Cunha try to stir the debate by studying the perceptions towards and appreciation of surveillance technology. One of the characteristics of securitisation is the growing use of information based on biometrics or the mechanisms to capture information in databases by registering physical and biological traces of the human body. Scholars in surveillance studies have largely discussed the implications of biometrics for human rights and have doubted its effectiveness in preventing and detecting crime. Up until now a limited number of empirical studies have tested these notions. Using a new DNA database that has been created in Portugal for criminal and civil identification purposes, the authors did empirical research in order to grasp the way DNA surveillance is perceived. They interviewed a wide range of social actors in the Portuguese context: Portuguese prison inmates (the surveilled), professionals in civil service in the domain of law, forensic genetics and policing (the surveillants) and finally academics in the fields of law and psychology. The perceptions of these people towards the use of DNA-technology for criminal investigation are sometimes unexpected.

Susanne Wigorts Ynvesson sets the tone for further debate about the moral issues of technology and surveillance specifically in the context of the workplace. Based on a literature review she asks some pertinent questions concerning surveillance in the modern work place. First of all, she examines whether there is a moral implication if employees are aware that they are being monitored or not. Secondly, she questions the blurring boundaries between work time and private time which makes it difficult to define where privacy begins and ends. Finally, the author deals with the complex question of how to protect privacy rules when using biometric information for identification purposes.

The previous articles all discussed security in the domain of crime and disorder. In the last text the authors Gudrun Vande Walle and Lieselot Bisschop discuss a case of human security and the consequences of translating a general social problem of ecological pollution into a security issue. The authors wonder if securitisation of ecological issues also causes the rather negative consequences of criminalisation, dispersion and violation of human rights. Or is securitisation of general life-threatening problems a blessing for the quality of life and the saving of the earth?

We thank the authors for contributing to this publication as part of the COST networking project 'Living in Surveillance Societies'. We also thank William Webster for his excellent and inspiring coordination of the COST-LiSS programme. Finally we thank COST for funding this project. We hope that the publication will stimulate the never ending debate about the effects of surveillance for the surveillant and the surveilled.

References

Haggerty K. & Ericson R. (Eds.) (2006). *The new politics of surveillance and visibility*. Toronto: University of Toronto Press.

Wood J. & Shearing C. (2007). *Imagining security*. Devon: Willan Publishing.

Beyond Criminal Law. On the Dutch anti-social behaviour agenda

*Patrick Van Calster and Marc Schuilenburg**

1 Introduction[1]

Problems in public space are a matter of widespread debate. Public space is synonymous to crime, filth, noise, traffic jams and juvenile delinquents hanging around street corners. One way the Western state tries to intervene in inner cities is based on the conviction that people are responsible for their own lives and should be held accountable. In other words, people should be responsibilised (Garland, 1996, 2001). Responsibilisation is rooted in the idea that the traditional criminal law approach is too limited to effectively combat crime in public space. Crime is a societal problem that needs to be addressed with the help of forces in society. From this perspective, the state wants to make the institutions closest to the population co-responsible for addressing and combating crime. In other words, combating crime is no longer solely a matter for the criminal law apparatus, but also one for schools, private citizens, volunteers, insurance companies, local administrations, hospitals and so forth. At the executive level, this has led to a wide range of new actors becoming part of the apparatus to detect security risks and maintain security in public space. In the ideal situation, successful responsibilisation should lead to a degree of self-reliance on the part of the parties who have been made responsible. However, although a great deal of research has been conducted on the efficiency and effectiveness of the criminal law system and its actors, very little attention has been devoted to the unique role new private actors have come to play in our immediate vicinity, the supervision they exercise, the techniques they use, the sentences they pronounce and the truths they adhere to. The question is whether the instrument of responsibilisation is indeed as legitimate and successful as the state seems to think.

---

* Patrick Van Calster is professor of Criminology at the University of Groningen, the Netherlands. Marc Schuilenburg teaches at the department of Criminal Law and Criminology of the VU University Amsterdam, the Netherlands.

1. Translation of this article was financed by the Translation Fund of the Royal Netherlands Academy of Arts and Sciences, and Reprographic Reproduction Rights Foundation (Stichting Reprorecht).

We would like to make a contribution towards this research and concentrate in this chapter on the Collective Shop Ban, a step taken in the Netherlands in an effort to make shopkeepers co-responsible for maintaining security. The Collective Shop Ban is an excellent example of a local agreement that serves as an instrument of social control, which Crawford refers to as 'contractual manageability' (Crawford, 2003). The new measure was introduced in the central business district of The Hague to prevent anti-social behaviour. A Shop Ban is enforced if a shopkeeper observes an individual exhibiting undesirable conduct. Undesirable conduct can refer to punishable acts as well as acts that are merely a nuisance. An example of the latter would be extremely ill-mannered behaviour towards salespeople. Depending on the severity of the conduct, a warning is first issued or a denial of entry is immediately imposed. The shopkeeper can decide whether to issue a warning or deny entry. A warning can only be turned into a denial of entry if the anti-social behaviour is repeated. The severity of the conduct determines how long entry is to be denied. Denial of entry can be imposed for six months or a year. During this period, the individual is not allowed to enter the particular shop or any of the other shops that belong to the Downtown Federation of Shopkeepers. So shoplifting at a supermarket can mean an individual can no longer go to the local pharmacy. If the order is violated, the individual is guilty of entering the premises illegally, which is punishable under Section 138 of the Dutch Penal Code.

In this article we examine how the Collective Shop Ban is applied in daily practice and how shopkeepers deal with their new responsibility. After all, in principle it is not mandatory for shopkeepers to enforce the Collective Shop Ban. They can do so as they see fit and take new steps of their own, which in turn can potentially influence the official policy. It is interesting to see whether they take advantage of this opportunity and how they do so. Ultimately, the measure is one with significant ramifications for those who are confronted with it. This is why we studied the attitudes and practices of the shopkeepers who impose the measure. We conducted 74 interviews with shopkeepers who carry out the Collective Shop Ban in The Hague. The shops are on Gedempte Gracht, Gravenstraat, Grote Markt, Grote Marktstraat, Haagsche Bluf, Hoogstraat, Spui, Turfmarkt, Venestraat, Vlamingstraat and Wagenstraat. The shops range from large concerns (Mediamarkt, Bijenkorf, Hema). to small shops. In addition, four interviews were conducted with city officials and seven interviews with private security guards. We also spoke to two people who work at the Downtown Federation of Shopkeepers, the actual policy-makers.[2]

2. The interviews were conducted by Jelmer Brouwer, Freek van den Engel, Joris van Heesch, David Knecht, Sophie Mommers, Desiré Peters, Lara van Roon, Moniek Weerts and Rolf van Wegberg. All the interviews were recorded and transcribed. Respondent anonymity was safeguarded. The authors would like to thank everyone for their time and effort.

2 The British example

Let us start by noting that steps like the Collective Shop Ban have not only been taken in the Netherlands. Comparable measures focusing on combating anti-social behaviour in public space have also been taken in various other countries. Perhaps the most well-known measure is the British *Anti-Social Behaviour Order*, passed to put an end to vandalism, littering, public drunkenness, noise and disturbing the peace. In addition, the *Crime and Disorder Act* enables the judge to enforce civil orders for the period of two years on minors above the age of ten. They are orders that forbid an individual to enter a certain area or associate with certain individuals during that period of time and thus keep the neighbourhood from being re-exposed to anti-social behaviour. Failure to obey an Anti-Social Behaviour Order is a punishable act. If the Crown Prosecution Service views this act as a summary offence, the case is brought before the Magistrates' Court. In a relatively simple procedure, the perpetrator can be given a fine and/or a prison sentence of six months. If the Prosecution Service views the act as an indictable offence, the Crown Court is authorised to look into it. In this much more formal procedure, a person who disobeys an Anti-Social Behaviour Order can be given a fine and/or a prison sentence of five years (Collins & Cattermole, 2006).

A study by Burney (2002) shows that Anti-Social Behaviour Orders are often requested and enforced to punish youngsters ostensibly heading in the wrong direction. In addition, the study reveals that the measure is used if there is no evidence to warrant the initiation of real criminal law proceedings or to grant an interim measure in anticipation of such proceedings (Burney, 2002, p. 477). Brown (2004). suggests that the authorities requesting the measure prefer exclusion to a short, and thus not very effective, prison sentence. A number of critical comments have since been made. In particular, the absence of due process guarantees has been denounced by the Commissioner of Human Rights of the Council of Europe (Gil-Robles, 2005).

Burney (2002) also notes that an Anti-Social Behaviour Order criminalises behaviour otherwise viewed as normal, such as hanging around at certain spots, so that the measure mainly targets certain individuals, especially youngsters. It seems as if Anti-Social Behaviour Orders are used more to combat crime and less to combat the anti-social behaviour that precedes it (Crawford, 2009). One reason for this is that it is easier to gather evidence of a violation of an Anti-Social Behaviour Order than to gather evidence in a criminal case. According to the Youth Justice Board (2006), this is why judges prefer to find someone guilty of violating an Anti-Social Behaviour Order. This is confirmed by day-to-day practice. Although Anti-Social Behaviour Orders are primarily formulated for anti-social behaviour on the part of individuals who have not committed any serious offences, the large majority of these orders are enforced against individuals who have been condemned for punishable acts in the past (Campbell, 2002).

It is clear from various product and case evaluations (Bullock & Jones, 2004). that although an Anti-Social Behaviour Order does reduce the repetition of anti-social behaviour, it does not influence the conduct of a large number of intractable youths. These evaluations do not, however, examine whether the approach simply makes youngsters go to a different neighbourhood, therefore the reduction in repeated anti-social behaviour may be attributable to this factor. In 2010 Theresa May, the British Minister of Home Affairs, announced the end of the Anti-Social Behaviour Order. Its transformation into an unintended badge of honour, in particular, played an important role in this decision. Some youngsters were not deterred by Anti-Social Behaviour Orders. On the contrary, it increased their social status within the own peer-group.

One might expect countries interested in passing measures of this kind to look to the British experience and learn from it. However, this has not been the case, at least not as far as the Netherlands is concerned. There is an overwhelming political consensus on civil measures to combat crime and anti-social behaviour. So let us now examine more closely the background of the new Collective Shop Ban and how shopkeepers are using it in The Hague.

3 Background of Collective Shop Ban

In The Hague three categories of conduct are viewed as warranting a Collective Shop Ban. The first pertains to petty offences, for instance shoplifting stealing an amount up to €50.00.[3] In the trial period applying to the measure, only a written warning about a Collective Shop Ban was issued for this conduct. This has since changed. According to the director of the Collective Shop Ban Enforcement Division, also speaking for the Downtown Federation of Shopkeepers closely involved in the introduction of the measure, most shops have also started imposing a Collective Shop Ban for petty offences. Warnings are sometimes given to minors. In 2007, 39 youngsters under the age of eighteen, one of whom was only ten, were given a warning. Once a warning has been given and the individual commits a new offence or exhibits anti-social behaviour at one of the shops it applies to, a Collective Shop Ban is imposed for a period of six months. The second category of conduct relates to more serious offences and can lead to a denial of entry for six months.[4] The most serious category of misconduct, which can also

3. Petty offences include anti-social behaviour, attempted fraud or forgery, attempted shoplifting, stealing an amount up to €50.00, insulting or threatening salespeople and/or shoppers without violence, destruction of property with damages under €100.00 and vandalism with damages under €100.00.
4. This category includes fraud or forgery involving an amount up to €250.00, theft of an amount up to €250.00, destruction of property with damages up to €200.00 and vandalism with damages up to €200.00.

include violence,[5] usually leads immediately to a denial of entry for a period of twelve months.

If a punishable act is witnessed by a security guard, salesperson or shopper, it is possible to search in an online database and see whether a warning or a Collective Shop Ban has been issued in the past to the individual in question. The security guard or shop manager can decide whether to issue a denial of entry. In addition, if a punishable act is committed, charges can be filed with the police. This is done in the event of a first offence or entering the premises illegally. In the framework of the project, the Public Prosecutor has expressed the intention to prosecute every single charge of shoplifting or disturbing the peace. However, according to the Police Department prevention advisor, this does not occur in all cases. Certainly as regards petty offences, it is not always advisable to file charges with the police, although not everyone shares this point of view. The director of the Collective Shop Ban Enforcement Division at the Downtown Federation of Shopkeepers does not concur with the Police Department prevention advisor and states, 'even if they steal a peanut, charges will be filed'.

Membership of a shopkeepers' association is a prerequisite for applying the Collective Shop Ban. The shopkeepers' association makes it possible to not only impose the denial of entry at one shop, but to do so simultaneously at all the member shops. Membership in a shopkeepers' association is not only a prerequisite for putting the measure into effect, it means the shopkeeper actively supports it. Collaboration implies certain, very specific, obligations. A protocol has been introduced in The Hague that participants need to conform to. The aims, which participants set for themselves in the protocol, are to improve security in the central business district, combat shoplifting and anti-social behaviour, emanate an overall preventive effect and discourage troublemakers. Not everyone voluntarily enters into this agreement. The director of the Enforcement Division at the Downtown Federation of Shopkeepers notes that certain shops in shopping centres are mandatory members of a shopkeepers' association and are thus automatically obliged to observe the Collective Shop Ban that applies to it.

The most important advantage to a shopkeeper is the reduction in income lost due to shoplifting. According to all the parties involved, this is one of the main effects of the policy. However, collaboration also has other advantages. Firstly, private security can be hired more effectively and often at a discount if the entire shopping street takes part in the Safe Business Hallmark. Secondly, the Centre for Crime Prevention and Security advertises an insurance discount at its site if the measure is put into effect. However, this discount has never actually been given, as noted by the Vice Chairman of the Downtown Federation at the consultations of the Downtown The Hague Board. The Association of Insurers has, however,

5. The third category includes fraud and forgery involving an amount above €250.00, theft of an amount above €250.00, insulting or threatening salespeople or shoppers using violence, manhandling personnel and/or shoppers and destruction of property with damages above €200.00.

promised to remit a one-time amount if the Safe Business Hallmark certificate is prolonged under the condition that this amount is spent on security facilities. Thirdly, the state offers subsidy options and participation can lead to tax advantages. Fourthly, in the event of anti-social behaviour the efforts of the city and the police are guaranteed. By not joining a shopkeepers' association, some shopkeepers are excluding themselves. The head of the City Centre Shopkeepers Association does 'not understand why anyone would not want to join in.'

4 How does it work in practice?

The difference and similarities to criminal law are crucial to a proper understanding of the Collective Shop Ban. Classic instruments of power, such as apprehending offenders and excluding them by denying entry, are used to impose this measure. The acts involved are traditional offences such as theft, fraud, defamation, intimidation and vandalism. These offences are traditionally included in the Penal Code. The powers of the Collective Shop Ban are, however, not stipulated in the Penal Code but in agreements or protocols locally accepted by the participating parties, in this case the Downtown Federation of Shopkeepers, the city, the police and the Public Prosecutor (Crawford, 2003). Investigations into these offences are made outside the Police Department and courts, namely by private parties, large-scale chains of shops and private security agencies. A criminal law suspicion in the sense of a *reasonable assumption of guilt*, as is stipulated in Section 27 of the Penal Code, is not required. To officially impose a denial of entry, a form needs to be signed by at least one witness and the offender. If offenders refuse to sign the form, a second witness can sign for them, thus making the document valid. This is also why legal rights are protected by a complaint procedure that not only offers the offender fewer guarantees than the ordinary criminal law system, it also asks offenders to take steps themselves to defend their case. Since these differences seem to go further than the similarities with classic criminal law, if only because the rules are not fixed in a universal code of laws that regulates all cases and all situations no matter what, and the police and courts are not in charge of the investigation of the offences, the Collective Shop Ban seems to be an excellent example of what we would like to refer to as *semi* or *quasi* criminal law.

Quasi criminal law does not work entirely outside ordinary or classic criminal law. In a sense, in fact, it works within the criminal law regime because in practice it requires a system within which the police and the Public Prosecutor take part. In the case of the Collective Shop Ban, the Downtown Federation of Shopkeepers has not only been given the support of the city authorities of The Hague, but also of two Public Prosecutors as observers. They concluded that, legally speaking, everything was fine. What is more, in the framework of the project, the Public Prosecutor has stated the intention to prosecute every charge of entering premises illegally as well as shoplifting. In addition, the Personal Information Protection Board and the Police Department have both approved the shopkeepers' policy. The police have a joint briefing three times a week where security staff members are informed about individuals with Collective Shop Bans. At these meetings photographs are shown of individuals who are denied entry to shops to make it easier to recognise them. Although the Downtown Federation of Shopkeepers would like to have been able to work with photographs, the Personal Information Protection Board has indicated that it is not allowed. This is why only certified security guards can see the photographs. The coordinator of the Collective Shop Ban project says the briefings are only useful for shops with security guards. Shops that have not hired a security company cannot attend the briefings. It is nonetheless clear that in exchange for safety and security in the central business district, the authorities have transferred their power to other actors (Wakefield , 2003), and thus to a different form of regulation.

Therefore, we see at least two levels of regulation in public space that are not only auxiliary to each other but actually intervene with each other. To the extent that the Collective Shop Ban expands and delineates the meaning of criminal law, quasi criminal law would seem to *supplement* it, to use a term coined by the French philosopher Derrida. In the French sense of the word, *suppléer*, supplement means to replace as well as to supplement. In the opinion of Derrida (1994), a supplement is not just any arbitrary addition, but a necessary one, like the space between the lines of this chapter, to an already existing reality or legal field. Since quasi criminal law is relatively new and its gradual expansion is supported by the authorities, we have examined exactly how shopkeepers use the Collective Shop Ban. We have focused on (a). how familiar the public is with the measure, (b). how arbitrary the enforcement is, and (c). whether there have been mutations of the measure, in other words whether individual modes of settlement have developed besides the ones described in the completed protocol.

(a). Shopkeeper familiarity with the measure

The quasi criminal law Collective Shop Ban undeniably serves a public interest. This means public interests are managed by private parties. In view of the responsibilisation process described above, this is not out of the ordinary. For a proper promotion of public interests, however, the powers of private actors need to be stipulated in writing and there needs to be public accountability. The formulation of the powers of the participating parties in agreements or protocols are part and parcel of the Collective Shop Ban in The Hague. To retain the quasi criminal law nature of the measure and properly put it into effect, it is important for shopkeepers to be well informed about the procedures so that they can be properly applied. Our research shows that the Collective Shop Ban coordinator, responsible for providing information about the procedure to be followed by the shopkeepers, has a hard time keeping them well informed. Shops close, and new ones open up. It is relatively easy for shopkeepers to join the Downtown Federation of Shopkeepers,[6] but just as easy for them to terminate their membership. As a result, the coordinator has only a very fragmented impression of what is going on in the work field. The incomplete list of members is not the only reason why it is so difficult to keep the shopkeepers informed. The shops themselves also play an important role. The Collective Shop Ban is often regulated via the main office of a chain of stores, but feedback to each separate shop is not always optimal. Many shops do have a participation sticker on their door, but the salespeople barely know what the sticker stands for. To be sure, the way shop managers are moved around can have an effect. They often do not stay long at one shop before they are transferred to another one. Furthermore, they often fail to pass on information about denials of entry to the managers who come and take their place. There are also shops that refuse to put the participation sticker on their door because, as a police official says, 'it can detract from the image of a chain of stores.' This suggests an inadequate sense of responsibility on the part of these shopkeepers.

It also appears that various shopkeepers do not know enough about how to use the Collective Shop Ban forms. They often do not know where they can request the forms, which is why they have not got any, and if they do have them, it is often unclear how the forms should be completed. A police official states that many of the forms are not filled out correctly. This means these denials of entry do not immediately go into effect and the forms first need to be corrected before they are legally valid. At the same time, it is striking that shops that do fill out the forms properly often comment on how undifferentiated they are. The security guards at a large chain store on the Grote Markststraat note, for example, that they always amend the forms because in all cases of anti-social behaviour a denial of entry of twelve months is otherwise enforced. The staff at another chain store on the same street note that the forms do not have enough space to add anything. This is why they always supplement the forms with their own reports. Another point made in the interviews is that many shopkeepers are unaware of the existence of a list of people denied entry to the shops. This database of denials of entry is managed by the Downtown Federation of Shopkeepers and can be consulted via the internet. If they do not know about it, shopkeepers cannot check and see whether individuals who exhibit anti-social behaviour have already been denied entry to the shops in the area, and cannot charge them with entering the premises illegally. According to the Downtown Federation of Shopkeepers, it is very easy to update the list every day on the internet, but the shops do need to have an internet connection on site. And that is not always the case. It is striking that the Downtown Federation of Shopkeepers says shopkeepers need to take responsibility for finding other ways of consulting the list, for example, by using their own computer at home or at a shop in the neighbour-

6. In order to take part in the policy, it is necessary to join the Downtown Federation of Shopkeepers.

hood that does have an internet connection. As is noted above, the Downtown Federation of Shopkeepers gives shopkeepers ample freedom to decide when and where they want to impose a Collective Shop Ban. The Downtown Federation of Shopkeepers has no objection to shopkeepers drawing up and adhering to their own criteria. This means one shop can be much stricter than another about imposing a Collective Shop Ban. We return to this below. In short, we see a relative unfamiliarity with the measure, with shops being given a great deal of freedom as regards the application of the measure. This takes us to the issue of arbitrariness in the enforcement of the measure.

(b). Arbitrariness in the enforcement

In order to provide for adequate implementation of the measure, agreements were made between the city, the police and the Public Prosecutor at the start of the Collective Shop Ban project in 2005 about what the shopkeepers can and should do. Not only does the protocol specify what shopkeepers should do if they observe a first offence or repetition, it states what the duties of the police and the Public Prosecutor in the cooperation framework are. However, although the project specifies a uniform plan of action for dealing with anti-social behaviour and shoplifting, in practice we see that this is difficult for shopkeepers to adhere to. Shopkeepers make any number of exceptions to the written agreement, which is one reason why, in practice, enforcement has come to be so arbitrary. Although there are often strong individual reasons for these exceptions, in general a number of practical reasons can be cited.

(1). Various shopkeepers say they simply do not have the time to impose a Collective Shop Ban. This is especially the case in the evenings when shops are open and it is very busy. Since there are so many customers on those evenings, it is virtually impossible for the salespeople to take shoplifters to the office and have them fill out the form. 'I cannot afford to do without one of my salespeople,' explains the manager of a shoe shop. Instead of going through the formal procedure with the offender, on these evenings he prefers to simply escort offenders out of the store. Various shopkeepers view this way of dealing with anti-social behaviour as easier and faster. Calling the police and filing charges take a lot of time. The manager at a pharmacy says: 'It takes too much time, I really would not even consider doing that.'

(2). Certain financial considerations play a role in the shopkeepers' decision whether to impose a Collective Shop Ban. Some of them note that the severity of the offence and the value of the stolen merchandise influence whether they impose a Collective Shop Ban. According to our interviews with security guards, one guard mentions the concrete example of catching someone stealing a magazine. In this case, charges are filed with the police for theft but no Collective Shop Ban is imposed because the security guard feels it is out of proportion. Not only can the relative lack of severity of the offence lead to exceptions, in concrete cases shopkeepers weigh the time loss against the value of the stolen merchandise. Especially as regards petty offences, a Collective Shop Ban is not viewed as worth the trouble. As one shopkeeper says, 'I am not going to go through the whole rigmarole with the police for someone who steals a five euro earplug.' The manager at a pharmacy on Turfmarkt says, 'What I earn in an hour is more than a twelve euro package of nappies.' This is why she does not always impose a Collective Shop Ban.

(3). Many shops do not have enough staff to effectively enforce the Collective Shop Ban. They are mainly one-man shops or at any rate very small ones.[7] Aside from the requirement to have enough salespeople on hand to deal with an offender separately, shopkeepers often believe it is not safe to fill out the Collective Shop Ban form with the offender. The manager at a clothing shop on Haagsche Bluf says, 'You don't just go sit down somewhere with a shoplifter, give him a form and ask him to please fill it out. And it really sounds weird if you ask him afterwards to please sit there and wait till the police come.' So small shopkeepers are more apt to just escort the offender out of the shop without making a case of it. Sometimes salespeople ask staff at other shops to come and help them. So what emerges is similar to Russian matryoshka dolls: cooperation frameworks within cooperation frameworks that keep getting smaller and smaller. Although the Collective Shop Ban is intended for the entire central business district, these microsystems come to the surface in random shopping streets or parts of the city. There is a framework of close cooperation like this at the shopping centre *De Passage*. As one shopkeeper recounts, 'Whenever there is a problem with someone who is aggressive, we call the neighbours. They come right over. Outside our neighbourhood, we don't have any contact with other shopkeepers about security.'

(4). Some shopkeepers appear to use double standards towards certain individuals. In fact one chain of stores has a staff member with a Collective Shop Ban himself. A security guard at an electronics shop made this surprising discovery. How did the shop in question respond to this discovery? 'Yes, but he stole from you. That does not mean he is also going to steal from us.' So here there is a unique situation of a shoplifter checking on shoplifters. This adds a new perspective to the assumption that a Collective Shop Ban always leads to the offender being denied entry to shops in the city centre. 'As long as you do your job well,' is how the respondents react to the fact that people are employed who themselves have a Collective Shop Ban. However, shopkeepers also note that specialised

7. This does not hold true for large chain stores such as Bijenkorf, Vroom & Dreesman, Mediamarkt and Kruidvat. They hire their own private security companies. Since these private security guards are there, there is no need for personnel to concern themselves with filling out forms or filing charges. But it is not worth the money for every shopkeeper to hire private security personnel. The consideration here is once again a matter of costs and benefits. As a pharmacy manager notes, 'We had a security guard here but I had to let him go because it simply cost too much.' Without a security guard, it is hard to enforce a Collective Shop Ban.

businesses 'need' certain types of customers. This is why steady customers who might have been issued a Collective Shop Ban at some other shop are still not denied entry. As long as they obey the rules at the shop in question, they can come in. Here again, there are underlying financial considerations on the part of the shopkeepers that motivate their conduct.

So the specific circumstances of each case and personal considerations play a role in whether a Collective Shop Ban is actually enforced. Shopkeepers who are rarely confronted with anti-social behaviour don't join shopkeepers' associations to take part in the Collective Shop Ban project. A sense of responsibility for what goes on in society seems to be linked in this connection to the shopkeepers' own business interests. From the inside, the system thus appears to be linked to the spot where it is in effect. This takes us to the question of whether or not there are mutations or deviations in the measures to be imposed.

(c). Individual modes of settlement

In classic criminal law, the punishment generally has to adhere to all kinds of rules of material and formal Penal law. This is necessary because penalties are imposed by the state for the violation of a specified norm. In an effort to reinforce legal security and judicial equality, the Public Prosecutor has formulated guidelines and judges use lists of amounts to be charged, so there is a certain uniformity in the nature and severity of punishments. As regards sentencing, the fixed law motivates the judge to make certain decisions. This motivation requirement, as is stipulated in Section 359 of the Penal Code, is there to protect the suspect from arbitrariness and inequality, especially if prison sentences are involved. The whole quasi criminal law nature of the Collective Shop Ban indicates, however, that due to the number of participating parties, varying from one-man shops to international chain stores with their own security services, individual considerations play a role in how the measure is enforced in practice. We have already noted the problems regarding the distribution of information about individuals who are denied entry. In addition, we have touched upon the practical problems shopkeepers are confronted with in imposing the Collective Shop Ban. However, other issues also play a role, for example, where individual settlements are concerned.

Sanctions in quasi criminal law are assessed within margins that are extremely wide. In the case of the Collective Shop Ban, the formal framework includes various categories of conduct. Yet some further modification should be added to the simple distinction between what is allowed and what is prohibited. Based on any number of individual initiatives regarding the settlement of violations, specific interventions have developed that no longer bear any relation to the fixed arsenal of general punishments (community service, fines, prison sentences and so forth) used by the classic criminal law system. It is clear from the interviews with shopkeepers that almost half of them have their own individual modes of settlement side by side, as it were, with the criminal law and quasi criminal law approaches. These vary from escorting offenders out of the shop to intimidating them. Although these settlements vary in how they are expressed, they are all based on shared sentiments and motivations and are generated in part by the ostensible ineffectiveness of the Collective Shop Ban. 'The very next day, there they are again,' says one of the interviewed shopkeepers about offenders. In the wake of the quasi criminal law denial of entry measure, four alternative modes of settlement regularly recur.

(1). Shopkeepers personally call the parents of the offenders, minors who shoplift or engage in anti-social behaviour. A number of shopkeepers feel this is extremely effective. With these youngsters, it is often only an isolated incident, they explain. Calling their parents makes enough of an impression on the youngsters that the police do not have to be called in. It has a corrective and normative effect, the respondents say. As a reaction to norm deviation, this mode of settlement contributes towards the power of the norm (Foucault, 1975). One might even say that shopkeepers are trying to protect the norm this way. In part, this explains why certain shopkeepers deviate from official sanctions and take measures that are apt to be more effective in the future.

(2). Some shopkeepers use humiliation or, more appropriately in relation to quasi criminal law, *naming and shaming*. Braithwaite (1989). notes that shaming is an important way of coping with deviant behaviour. It is thought to have a deterrent effect. Braithwaite draws a distinction, however, between two types of shaming, i.e. stigmatising and re-integrative shaming. Stigmatising shaming emphasises the offender's criminal activities and rejects the person as well as the conduct. Braithwaite feels this form of shaming is dangerous because it destroys the link between the offender and society, probably for the rest of their lives. As a result, the offender identifies with the label of criminal and may end up playing the role without ever stopping. Re-integrative shaming is much more positive and does not focus on the person but tries instead to rehabilitate the offender. The aim is to restore the link between the offender and society. Our research seems to suggest that in the enforcement of the Collective Shop Ban, it is mainly stigmatising shaming that comes to the fore. One shopkeeper on Haagsche Bluf notes that she deliberately gave an offender negative publicity. 'We worked with the police and printed his photograph on a large poster with the word Wanted and hung it up on the window.' Another way of afflicting shame is by publicly apprehending offenders and then having the police come and take them out of the shop through the front door.

(3). Shopkeepers use their own fines as an alternative for the punishment stipulated in the protocol. It is generally assumed that in addition to the punishment, the deterrent effect of a sanction depends on the chances of getting caught, the speed of sentencing, and the certainty that a sanction will actually be carried out. The owner of a discount store recalls the choice a thief was given. 'Either you pay twice the price or I'll call the police.' It is mainly petty offences that are involved, like stealing sweets. In other cases, offenders are only reprimanded for their behaviour. Someone caught stealing is told to give back the stolen merchandise and leave the store.

Lastly, we would like to say something about the Rogues Gallery drawn up to solve the problem of recognising individuals with a Collective Shop Ban. The list the shops are given only contains names and no photographs of the individuals so there is a problem recognising them. 'We take pictures of people who steal,' says the head of security at a shop on Grote Marktstraat. 'Our communications room is full of photographs.' All the pictures are kept in a Rogues Gallery, a book for salespeople so that they can recognise shoplifters. Various large chains of stores with their own security work with a similar system. Photographs of individuals who are denied entry are internally available to security guards, but not to salespeople. Although the authorities say the infrequent charges of entering the premises illegally prove that the Collective Shop Ban is working, the shopkeepers believe the infrequent charges mainly show that the policy do not effectively stop repeat offenders. This option does not seem to be operating completely effectively yet either.

5 Conclusion

One of the unique features of a city is the way it promotes shopping. The city is the primal landscape of consumption (Hayward, 2004). This opens up opportunities, but it also has repressive consequences. New measures in the area of public order and security make the latter an everyday occurrence. The debate on anti-social behaviour and problems of crime in our urban environments is but the latest in a long tradition of public disquiet over the urban condition. As a result, over the past twenty years we have seen the rise of a new kind of security management in the city. The state, and consequently the police, are now only two of a motley collection of actors active in this area. It has become a matter of course that, for instance, shop owners too take responsibility and develop and implement their own security programmes. However, the governance of security is a more complex phenomenon than is usually assumed and described. Our understanding of the causes and correlates of crime is substantial, but comparatively little is known about the management of security within quasi-public spaces.

In highlighting some results of the interviews with retailers, we have shown that the Collective Shop Ban is applied to various degrees and with an ample dose of arbitrariness. Personal convictions play a major role here. One shopkeeper might impose the strictest version of the Collective Shop Ban without any hesitation, whereas another might be much more lenient. As a result, there are all kinds of mutations of the measure. We discovered, for example, the existence of a process of stigmatising shaming. Criminologists like Braithwaite (1989). cautioned us more than twenty years ago about this kind of process because it can significantly disturb the relation between the perpetrator and society, causing the perpetrator to identify even more strongly with the role of a criminal and develop a criminal career.

Our findings do not lead us to categorically reject this development towards quasi criminal law measures. Personal considerations, for instance, play a role in every form of regulation and can lead to a far more satisfactory final result in individual cases than strict adherence to guidelines. Instead, the results of the interviews lead us to conclude that there is something of a double order in this system of quasi criminal law. On the one hand, there is the Collective Shop Ban officially formulated in a protocol that has been agreed to by the local city authorities, Police Department and Public Prosecutor. At an organisational level, this results in policy-makers claiming the success of the Collective Shop Ban on the basis of evaluations and the number of denials of entry imposed. In 2009 for example the Main Retail Trade Association announced that shoplifting could be reduced to 60% as a result of the measure. More than 1,500 denials of entry have been imposed by the shopkeepers of The Hague. As a result of its reputed success, more and more cities are introducing the measure and expanding it to include cinema or tram bans, for instance. Nowadays Amsterdam, Apeldoorn, Arnhem, Beverwijk, Den Helder, Deventer, Eindhoven, Gouda, Heerlen, Helmond, Leeuwarden, Leiden, Leidschendam, Rotterdam and Utrecht have either introduced this policy or have advanced-stage plans to do so.

On the other hand, there is the level of interaction between the participants that adds new openings and creates different links than those at the structural level. Here we see that within the area concerned, the official points of departure are disturbed, for example, as regards the distribution of information, arbitrariness in enforcement, and mutations in measures that have developed. At this level it is much more difficult to draw borders. This presents questions as to the wisdom of the state uncritically setting a responsibilisation process in motion. It raises the question of whether dealing with offences like theft, fraud, defamation, intimidation and vandalism can easily be left to private actors such as shopkeepers. After all, not only has the quasi criminal law Collective Shop Ban been invented by shopkeepers, it is these very same people who also enforce it. This makes the borderline between the individual private realm and the public realm a very diffuse one. After all, whose interest is served by forgoing certain constitutional rights, such as adequate legal protection, the citizen or the shopkeeper?

References

Braithwaite, J. (1989). *Crime, shame and reintegration*. New York: CUP.

Brown, A.P. (2004). Anti-Social Behaviour, Crime Control and Social Control, *Howard Journal, 43*(2), 203-211.

Bullock, K. & Jones, B. (2004). *Acceptable behaviour contracts addressing antisocial behaviour in the London borough of Islington*. London: Home Office.

Burney, E. (2002). Talking Tough, Acting Coy: What Happened to the Anti-Social Behaviour Order? *Howard Journal, 41*(5),469-484.

Campbell, S. (2002). *A review of Anti-Social Behaviour Orders*. London: Home Office.

Collins, S. & Cattermole, R. (2006). *Anti-social Behaviour and Disorder. Powers and Remedies*. London: Sweet & Maxwell.

Crawford, A. (2003). Contractual governance of deviant behavior, *Journal of Law and Society, 30*(4), 479-505.

Crawford, A. (2009). Governing through anti-social behaviour. Regulatory challenges to criminal justice' *The British Journal of Criminology, 49*(6), 810-831.

Derrida, J. (1994). *Force de loi. Le Fondement mystique de l'autorité*. Paris: Galilée.

Foucault, M. (1975). *Surveiller et punir. Naissance de la prison*. Paris: Gallimard.

Garland, D. (2001). *The Culture of Control. Crime and Social Order In Contemporary Society*. Chicago: Chicago University Press.

Garland, D. (1996). The limits of the sovereign state. Strategies of crime control in contemporary society, *The British Journal of Criminology, 36*(4), 445-466.

Gil-Robles, A. (2005). *Report by the Commissioner for Human Rights on His Visit to the United Kingdom*. Council of Europe: Office of the Commissioner for Human Rights.

Hayward, K.J. (2004). *City Limits. Crime, Consumer Culture and the Urban Experience*. London: Glass House Press.

Home Office (2004a). *Defining and measuring anti-social behaviour*. Home Office development and practice report no. 26. London: Home Office.

Home Office (2004b). *Perceptions and experiences of antisocial behaviour*. London: Home Office.

Jones, T. & Newburn, T. (1998). *Private Security and Public Policing*. Oxford: Clarendon Press.

Youth Justice Board (2006). *A research into Anti-Social Behaviour Orders given to young people between January 2004 and January 2005*. London: YJB.

Wakefield, A. (2003). *Selling Security: The Private Policing of Public Space*. Cullompton: Willan.

Surveillance in the supermarket: Technology and the pluralisation of crime control

*Francisca Grommé**

1 Introduction

In 2007, a city in the Netherlands provided its shop owners with a facial recognition system to detect known shoplifters. The local government presented the project as a favour to the retail sector in the area to enable entrepreneurs to combat retail crime more actively and more resolutely. This smart camera system compared faces of individuals in the crowd with police photos of known local shoplifters. The shop owner was alerted when a match was established between a person having entered the shop and a database photo. The project started out with a one-year pilot in one supermarket to learn if and how the technology could be used for this purpose. In spite of the city's high expectations of this technology, the supermarket employees were critical of the favour presented by their local government: would a database with police photos contribute to their own knowledge? And would it make their work easier? Moreover, the retail sector in this city had no ambitions to increase its involvement in policing activities, despite local government encouragement. Given this discussion about the position of retail in crime control, the introduction of facial recognition in the supermarket raises questions about the use of technology to increase retail involvement in crime surveillance.

In this paper, I ask how facial recognition intervenes in the relations between the various actors involved in monitoring and controlling shop crime. In particular, I focus on the way in which technology affects the pluralisation of surveillance, understood here as the state's effort to mobilise new types of actors. State agencies no longer regard themselves as the main providers of security, but actively and openly seek the role of facilitators or partners (Rose, 2000). Consequently, the state should be understood as only one of the nodes in an extended crime control network that includes a variety of public and private agencies, relationships, programmes and techniques (Den Boer, 2004; Loader, 2000; Rose, 2000; Van Steden, 2007). The involvement of both state and non-state actors in crime

* Francisca Grommé is a PhD candidate at the Amsterdam Institute of Social Science Research (AISSR) at the University of Amsterdam (Department of Anthropology). She does research on pilots with surveillance technologies in everyday practices of crime control. E-mail: f.gromme@uva.nl.

control is not a new phenomenon (consider, for example, Rembrandt van Rijn's *The Night Watch*).[1] However, the roles of private actors in crime control, including security agencies, retailers and technology suppliers, have nowadays become more formalised (Van Steden, 2007).

The pluralisation of surveillance in crime control is a significant development because, as is often claimed, it can result in more 'intense' and 'ubiquitous' forms of monitoring and social control (Lyon, 2007; Marx, 2002). In the Netherlands, various developments in the domain of surveillance and crime control have led to concerns about the infringement of new crime control measures on citizens' everyday lives. Since the 1980s Dutch crime policy has increasingly prioritised security, safety, efficiency and service provision over privacy and proportionality. Crime policy has become more risk-driven, aiming for prevention, while a strict application of criminal law has replaced a tolerant approach (Boutellier, 2002; Koops, 2011). At the same time, the transfer of state responsibilities in crime control has become an explicit policy aim (Van Steden, 2007; Van Stokkom, 2009). As for the use of new technology in the field of policing, Bert-Jaap Koops (2011) notes that a supply-driven line of thinking has become prevalent. Concerns about privacy, proportionality and legislation are now set aside to enable the use of new technology in crime control. Thus, a technological advantage over criminals is given priority over integrating new technology into law and practice.

Despite increased attention to these issues, we know little about the interaction between the pluralisation of policing and the introduction of new technology at specific sites. Recent scholarly work about surveillance suggests that technologies play an important role in this process because they link different types of actors in 'networks' or 'complexes' of surveillance (Introna & Wood, 2004; Prainsack & Toom, 2010). Building on Gilles Deleuze and Félix Guatarri's (1987) notion of the rhizome, Kevin Haggerty and Richard Ericson (2000) have put forward one of the most prominent conceptualisations of the form and growth of surveillance: the surveillant assemblage. The metaphor of the rhizome refers to the root structure of weeds that grow independent of the main roots, but may shoot up in different locations and grow without a central order or hierarchy. In line with this notion, Haggerty and Ericson suggest that the surveillant assemblage is a heterogeneous gathering of loosely connected actors, including people, institutions, technologies and knowledge. This assemblage can be considered as a unity to the extent that the actors may work together as a 'visualising device' for a target population.

In contrast to the idea of disciplinary surveillance (Foucault, 1995), this line of theory does not consider surveillance systems as discrete, institutional entities. Surveillance assemblages are expansive; they 'grow like weeds' (Haggerty & Ericson, 2000, p. 614) in the sense that they link ever more actors in fluid configurations. Haggerty and Ericson suggest that surveillance technologies play an impor-

1. As Ronald van Steden (2007) points out, the famous seventeenth-century painting *The Night Watch* by Rembrandt van Rijn portrays a group of guards that operated independently from the state.

tant role in the enrolment of new actors. Technologies participate in translating the subject of surveillance into disembodied fragments, the 'data doubles'. By doing so, they create new target populations that may link actors in the surveillant assemblage. In addition, technologies make connections by integrating databases, for example, when combining consumer databases, social security databases and police information systems to create a profile for a specific individual. These technologies do not necessarily need to be 'high-tech' as, according to Ericson and Haggerty's line of argumentation, paper files may have the same effect.

Missing from this account is a comprehensive consideration of the ways in which surveillance technologies are implemented in their environment in order to observe a target population. The assumption is that ready-made artefacts are implemented in a new context as a matter of course. In practice, however, they need to be integrated within the relations, strategies, techniques, norms and preferences at a specific site; a process that can involve contingency and resistance (Akrich, 1992; Callon, 1986b). Only limited attention is paid to discrepancies and conflict among the actors, partly because the actors are granted limited agency. This is to say that in assemblage theory the actors have no capacity to change the world around them. Instead, agency is attributed to the theoretical concept of the assemblage itself (Prainsack & Toom, 2010). Kelly Gates (2010) demonstrates that the effects of new technologies are not self-evident in a case study on a facial recognition pilot in Florida. Here, the technology served as a site of struggle over police competences in crime control. Moreover, ethnographic research shows that those who perform surveillance can be reluctant to accept more surveillance tasks, for example, because they prefer to function as hosts or workplace coordinators (Helten & Fischer, 2004), or because they experience monitoring activities as physically tiring (Smith, 2004).

In this paper, I use the notion of 'situated surveillance' to attend to the reconfigurations and contingencies that accompany the introduction of surveillance technologies. Christopher Gad and Peter Lauritsen (2009) suggest that surveillance is an event that only succeeds with various devices, such as paper files and computer programmes, act together. Consequently, 'the surveillant' is not a single operator, but a heterogeneous combination of humans, artefacts and sources of information at a specific time and place (M'charek, 2008). Introducing a new technology, such as facial recognition, can involve a reordering of the relations between the actors that jointly operate to visualise a target population.

In line with Gad and Lauritsen's proposed approach of 'theoretical agnosticism',[2] my aim is not to develop a generalised argument for a model of the form or structure of surveillance, such as the surveillant assemblage or the Panopticon (Foucault, 1995). Instead, I intend to provide an in-depth analysis of the ways in which technologies are involved in the pluralisation of surveillance in the super-

2. Gad and Lauritsen have adopted the original concept from Michel Callon (1986a) to denote a mode of analysis in which a search for general theory is replaced by a renewed focus on practices.

market case. By studying surveillance as a situated practice of making shoplifters visible, I show how facial recognition was, in this case, part of an effort to further enlist the supermarket in monitoring petty crime. I furthermore address the contingencies of this process by showing that the supermarket was not mobilised as planned due to an inability to integrate the technology with the local surveillance practices. While surveillance technologies can potentially engage new actors, they do not necessarily do so because they may refuse to take part in a particular mode of surveillance. In other words, not only subjects of surveillance resist, but resistance may also exist within the networks that perform surveillance.

2 Surveillance as a situated practice

Technologies are not passive entities in surveillance. As Amade M'charek (2008) argues with regard to the practice of making suspects visible on the basis of DNA, technologies may, depending on the context, lead to a reorganisation of the relations between actors. In a similar vein, facial recognition was designed to interfere with the existing relations between objects and people that were part of the strategies and routines of monitoring shoplifting in the supermarket.

In order to show how facial recognition intervened in surveillance practices on the work floor, I approach surveillance as a situated, collective practice that introduces only partial 'visions'. In her argument for the partial perspective as a basis for a feminist practice of science, Donna Haraway (1988) proposes the metaphor of vision to refer to partial observations of the world that are mediated by bodies, instruments, culture and politics. When applied to the observations made in surveillance control rooms, it becomes clear that what surveillants see are not straightforward representations of the world out there. Instead, they constitute new, hybrid subjects (Haggerty, 2006). Consequently, making observations does not entail the creation of an overview of everything from nowhere, or doing 'a god trick' (Haraway, 1988, p. 581). In fact, in most control rooms operators make detailed observations by seeing as little as possible because this is the only way to understand and manage a complex world. In their study of 'oligopticons' in Paris, Bruno Latour and Emile Hermant (2006) show how observatories, marketplace coordinators, and police control rooms exclude most of the world in order to observe what is relevant to them.

Surveillance practices are easily disturbed because the various devices involved in making something visible may not act together (Dubbeld, 2005; Gad & Lauritsen, 2009). For example, Gad and Lauritsen show how fishermen might not be detected by radar because they found a way to block the signal. This is not to say that successful observations depend entirely on 'lucky circumstances' in which everything comes together. The notion of situatedness refers to the ways in which surveillance is embedded in specific artefacts, contexts and routines. In crime surveillance, visions are often constituted by professional groups. It is a skilled practice drawing on collective experiences, the trained use of artefacts, peer-to-peer negotiations, hierarchical relations, narratives, professional identities and tacit knowledge (Grasseni, 2007). As Cristina Grasseni suggests, these practices are part of the negotiations that take place between standardisation (for example, Latour, 1990) and individual subjectivities (for example, Norris & Armstrong, 1999).

In the following I study surveillance as the professional practices in which various contextual logics are performed (Mol, 2008). The logics, or 'rationales' (Mol, 2008, p. 8), of practice include object definitions and strategies to handle that object (Mol & Berg, 1994). Importantly, logics, strategies and objects emerge from practices in which artefacts and humans take part and consequently do not operate as 'top down' power mechanisms (Law, 1992). To examine the logics of supermarket surveillance in this paper, I focus on crime control strategies and target group definitions. I start out with an account of the contextual logics of supermarket surveillance before the start of the facial recognition pilot. Next, I show how facial recognition was meant to reorder surveillance practices to perform a different logic of surveillance that included the mobilisation of the supermarket to monitor a more broadly defined target population. Finally, I argue that in this case, the new artefact did not lead to the introduction of this ideal logic.

3 The logics of supermarket surveillance: Collecting strikes and care

The context: Anarchy in the supermarket

In the supermarket that is the subject of this case study,[3] the employees had adapted their surveillance strategies and routines to their local situation. The supermarket was generally thought of as a problematic site, where a disproportionate amount of stealing took place. This supermarket was located in the city centre, where it welcomed tourists, business people and other passing visitors, as well as local residents. The employees felt they had little control over theft because of the large size of the store, the high number of visitors and the high racks that blocked their lines of vision in the store. A team of eight private security

3. In order to protect the identity of my informants, I refrain from mentioning the location of the pilot and the names of the organisations and persons that I interviewed for this paper. Moreover, citing the full titles of some policy documents and internal reports would reveal the names of the location and the organisations involved. In these cases, I do not cite the full reference in the bibliography. Instead, I refer to the type of document in the footnotes.

officers, working in shifts of two or three, were tasked with keeping order in the store.[4]

The supermarket was surrounded by a square known to local police officers as a site where drug dealers mixed with addicts, homeless people, tourists and locals. The security officers and the supermarket's local management regarded the group of people that often spent their days on this square as the supermarket's main source of nuisance. Commonly referred to by local management and employees as 'the community' or 'the culture out there',[5] this group was perceived as a constant threat to the supermarket. Accordingly, the official task of the guards was to 'prevent a state of anarchy in the supermarket'.[6] The security officers were instructed to prevent and detect shop crime in addition to their general task of making the employees and customers feel safer. In 2004 the supermarket implemented a banning order policy, comparable to the Anti-Social Behaviour Orders in the UK (see Donoghue, 2008). The banning order is essentially a means of local governments to reduce police intervention, because it allows the store's employees to impose the banning order themselves, provided they call in the police when they have arrested a suspect for theft or nuisance.[7] Individuals that break the house rules can be banned from the store for a year. Among this supermarket's house rules were: a prohibition of theft, harassment and substance use, but also the obligatory use of a shopping cart, a prohibition of eating inside the store and the obligation to follow staff instructions.[8] If a person with a banning order should attempt to enter the store again and refuse to leave, he or she could be arrested for trespassing domestic premises.[9] The banning order policy of this supermarket was very strict; every case of shoplifting or misdemeanour was

4. This paper is based on fifteen in-depth interviews and twenty-eight hours of observation between April 2006 and February 2007. Also, I had access to project plans and reports. This information was complemented with policy documents and the minutes of city council meetings on this pilot. The interviews include conversations with all members of the project group that designed and executed the pilot (police, public prosecutor, local government, technology consultant); the local manager of the supermarket where the pilot was performed; the supermarket's higher management; and city councillors. The average length of the interviews was 90 minutes.
The observations include three visits to technology tests (twelve hours in total) at the supermarket, and two visits to the control room of the private security officers (two working days or sixteen hours). My aim was to gain an understanding of the security officers' routines by interacting with them in a setting they felt comfortable in and where they were the experts as opposed to a more formal interview situation. Being in the control room also provided me with an insight into the various documents and tools the security officers use, and into the space they operate in.
5. All quotes are translated from Dutch by the author, unless otherwise indicated.
6. Work manual private security guards, 2004 (internal document).
7. In 2007, shop banning orders were only valid for individual supermarkets. At the time, this city had not yet introduced the collective banning order. For an analysis of the collective banning order, see Schuilenburg and Van Calster (2009).
8. Work manual private security guards, 2004 (internal document).
9. Dutch Code of Criminal Procedure, Article 53. The supermarket is privately owned and managed. Therefore, unlawful entry is interpreted as trespass on *domestic* premises.

followed by a banning order, even if the item stolen was only worth 50 cents.[10] For the supermarket the issue was not the value of the stolen products but the repeated acts of shoplifting by some individuals at this location. The local manager expressed the problem of petty theft as follows: 'The incidents themselves are small, but what is small if you do it all the time? It becomes serious, like somebody kicking you in the shins a thousand times.'

The logic behind this strict policy was that known shoplifters 'collect strikes'. Local management and the supermarket's private security officers had set out to arrest 'their' thieves as often as they could, each arrest resulting in a strike. According to a national act that was passed at the time (the ISD Act), the more often a person is arrested and taken to the police station, the more likely it is that this person will be detained for a longer period of time.[11] As a consequence, individuals that repeatedly commit minor offences can be detained for two years, whereas before they could only be 'kept off the streets' for a short period of time. In the supermarket, the shop ban was interpreted within the framework of the ISD Act. Collecting strikes within this framework was thought of by the supermarket's local management and security employees as the only effective method of diminishing crime and harassment. In practice, this approach meant that the primary goal of the security officers was to catch as many thieves as often as possible, as opposed to preventing theft.

The collective work of detecting thieves

The logic of collecting strikes was performed by the objects, knowledge, regulations and people in the supermarket in order to detect and arrest particular types of shoplifters. In this section I will take a closer look at the collective that performed this logic and the target groups that were watched. The primary focus of supermarket surveillance was to catch people in the act of stealing. In the supermarket, theft was only a fact when an individual passed the cash register without paying for an item. The security officers spent most of their days studying CCTV images. When the guards found a potential shoplifter, they followed their suspect's actions with several cameras. If they observed a person stealing, they followed this person until he or she had walked past the cash register, which allowed them to record the theft on videotape. Only then did they exit the control room, conveniently placed next to the cash register, and arrest

10. Official reports on the number of banning orders that the supermarket imposed do not exist. The supermarket security guards and the manager estimated that one or two persons were arrested every day and police arrests for disturbance of domestic peace were carried out about five times a month. In contrast, a supermarket in a small town issued only one or two banning orders annually. The current arrest rate is, however, still a vast improvement on the estimated figure of forty arrests in one week before the implementation of the new banning order policy.
11. ISD Act (*maatregel Inrichting Stelselmatige Daders*: Custodial Order for Repeat Offenders Act). Dutch Code of Criminal Procedure, Article 38m through 38u.

the suspect. The supposed offender was then taken to a tiny cell next to the control room, where the security employees photographed the suspect, registered his or her details and issued a banning order.

Various objects played a part in detecting theft. The store used 32 cameras, of which seven belonged to a CCTV network that allowed for manually zooming in and out and changing the angle of view. Black-and-white photos of known offenders were attached to the wall. These photos showed individuals that were banned from the store, and were made by the private security officers using mobile phones. Underneath the photos, the guards had made handwritten notes of their names and possible risks (aggressive behaviour, for instance). In addition, fifteen colour photos showed the most notorious local offenders as defined by the police. These photos were given to the guards by the local police station. Furthermore, the security officers used paper files with information about the banned individuals and the circumstances in which they were banned. These files were used for longer than the legal term of one year in case the guards needed information about persons that had received a banning order more than a year previously.

Monitoring was done partly by the acquired skill of swiftly switching between the screens and by changing the camera's angle with a joystick. In addition, the security officers used their professional knowledge to detect suspect behaviour. In general, the practices of the guards focused on two target populations, each requiring a separate set of skills, knowledge and devices to be identified. First, the guards tried to detect 'normal people that steal', or the Shoplifters/Normal persons. The security officers acted on the principle that 'everybody steals': a thief does not necessarily look like a thief. This was a difficult task, and consumed most of the security officers' working hours. They were taught what to look for by their colleagues and through their professional training. While operating the cameras, the guards searched for suspicious behaviour, such as nervousness and hesitation. In addition, they paid attention to a number of markers, such as worn out clothing or a 'tourist outfit'. No strict guidelines existed, however. The guards not only focused on individuals, but also paid close attention to the goods on the shelves. If a large stack of products suddenly went missing, this was a clue to search for suspect activity. The security employees made sure the cameras were not blocked, were pointed at the goods from the correct angle and that the shelf stackers continuously replenished the shelves. As a ground rule, the expensive products had to be placed at strategic places far away from the exit of the store and in view of the cameras.

Second, the guards kept an eye out for familiar faces, or the Shoplifters/Risky individuals[12]: those with a banning order and members of 'the community' that often loitered on the square just in front of the supermarket but were not necessarily banned from the shop. As the guards mentioned, their interaction with this group resembled playing 'cat and mouse games'. The guards pointed these individuals out to each other and they observed them from the entrance to the store, occasionally having a chat with them. Because the guards worked in other supermarkets in the city, they also knew about the activities of this group in other places. This interaction worked both ways, however: the men and women in this group had supposedly also learned about the routines of the security officers and used this knowledge in their attempts at shoplifting.

Besides being part of a logic of collecting strikes, the target group of Shoplifters/Risky individuals was important for the enactment of a second logic of surveillance of the security guards: care. At this location, the guards frequently experienced aggression and threats. To prevent personal injury, they used photos, notes about earlier behaviour, outdated files and stories as resources of information. Updating each other on the group of Shoplifters/Risky individuals was essential because, in the guards' experience, the appearances of people that live on the streets could change fast, e.g. they could gain or lose weight quickly.

Interaction with the police

To summarise, the supermarket's surveillant was a hybrid of the security guards, their knowledge of risky individuals and suspicious behaviour, cameras, paper files, photos and legislation. In addition, the guards organised the physical environment of the supermarket to ensure maximum visibility. They used their resources in line with the target populations they intended to identify according to the logics of collecting strikes and care. With regard to the relation between the supermarket security team and the police, police interference in determining the target population was limited to the supply of fifteen colour photos of police-registered shoplifters.

The police, however, wanted to cooperate with the security guards more closely. Privacy regulations did not allow for sharing more police information with the supermarket, but local police officers were eager to expand the guards' activities to a wider target group of frequent offenders in the area around the supermarket. The police tried to motivate the security guards to keep an eye on 'the community' on the square and to report suspect activity to the police. At the same time, the police attempted to limit the guards' activities with regard to policing, repres-

12. I use the term 'risky individuals' to describe my informants' target groups as accurately as possible. By using this term I do not intend to evoke Nikolas Rose's more general and analytical notion of 'risky individuals' as those people who are excluded by risk-based techniques in advanced liberal societies (see Rose, 2000).

sive action and the use of methods beyond their competences, such as the use of threat. Instead, their aim was to make sure the security officers complied with government guidelines on the use of the banning order, which included only limited monitoring, preventive and proactive behaviour, approaching a suspect *before* he or she walks past the cash register and a minimum use of threat.[13]

In contrast, the security officers preferred the opposite situation: they rejected police interference with their target populations and operated with the grey areas of what was permitted for private security officials. With regard to their target group, the guards defined the problem of shop crime as 'the outside coming in': problems in the world outside the supermarket, like drug addiction, were considered causes of shoplifting and nuisance. The security officers regarded these outside problems as police responsibilities and demarcated their area of activity at the entrance of the supermarket. At most, they watched the square from the doorstep. On busy days, they would drench part of the square with water to make sure that people would not 'hang out' there.

As for their methods of surveillance and policing, the supermarket security guards used the undefined rules of working in what they called a 'semi-public space' to their advantage. Not being tied to a strictly enforced code of conduct, as the police are, gave them a degree of freedom in tackling the diverse, and sometimes legally undefined, problem of shop crime. For example, issuing banning orders required that the security employees take and store photographs of the arrestee.[14] To do this required the suspect's permission. The guards would trick the arrestee into giving permission by using specious arguments ('otherwise we will call in the police' – which they would do regardless). Furthermore, the security guards were not allowed to search bags, but according to their own accounts, they would sometimes use threat to obtain permission. Thus, in terms of target groups and surveillance strategies, the guards refused to become an 'extension of the police'.

4 Introducing facial recognition

Designing the system: One store, two databases

Facial recognition introduced an automated method to detect shoplifters. During the pilot, camera-captured, real-time images of the people entering the store were compared to a database with mug shots of individuals that had received a banning order from the supermarket during the past year. The basic principle of this technology was feature extraction: the measurement of the distances between

13. Police information for local retail, police document, 2007.

14. Police information for local retail, police document, 2007.

various facial points, for example, the distance between the eyes.[15] The technology established a match between the algorithm that expressed the characteristics of the entrant's face and an algorithm in the database.[16] This was, and is, a challenging operation in an uncontrolled environment where people might suddenly turn their faces and where light conditions vary (Philips et al., 2005). Therefore, a match can never be established with a 100 per cent certainty. In the supermarket, the match that was statistically most likely to be correct was presented to the supermarket employees.

In the eyes of the city and regional police management, facial recognition addressed a complaint about the banning order that the city was often confronted with: retailers simply had no time to be on the lookout for people with a banning order, and were unable to recognise them after a few months had passed. But this project not only offered a tool to alert shop owners to the presence of the people to whom they had issued banning orders. The facial recognition system was also connected to a second database of police-registered frequent offenders.[17] The police database contained photos of individuals who had been repeatedly arrested for minor offences during the past years. It was composed of 1,250 frequent offenders in the region who had been convicted for shoplifting at least once, in some cases in addition to other offences.[18] Some of these men and women might have been registered by both the shop and the police, but a check for double entries was not made. This second database was exclusively police property. The hard disk was situated at the police station, and could not be searched by supermarket staff. Only when the system established a match between an individual in the supermarket and a police-registered offender would the photo and name be revealed to the supermarket employee on a hand-held palmtop.

15. Feasibility study of the project, technology consultant, 2005 (internal document).
16. The significance of the match is based on a comparison with a mean algorithm of the population. For a description of the status and operation of this technology at this time, see Introna and Wood (2004).
17. Final draft of the project plan, technology consultant, 2005 (internal document).
18. Different definitions of 'frequent offender' circulate in policing practices in the Netherlands. In its most general definition, a frequent offender is a person that commits offences on a regular basis, ranging from non-violent offences to violent robbery. This group of offenders is thought to have a disproportional influence on crime rates. In this case, the minimum threshold for the database was one police record of shoplifting during the previous twelve months and two police records during the individual's life. The information was retrieved from the EDISON (*Electronisch documentatie informatie systeem voor opsporingnetwerk*: Electronic documentation system for investigation network) police database. It should also be noted here that, like any category or label, the category of 'frequent offender' can be highly problematic. Here I use the terminology that I encountered during my fieldwork. Although opening up this category (and indeed other categories used in crime control) for discussion would be a highly legitimate project, it is beyond the scope of this paper.

As a result, new target groups were introduced to the supermarket. The use of police information by private actors was presented as the main innovative feature of this project. Alongside its presumed usefulness on the shop floor, the inclusion of a frequent offender database legitimated the city's financial investment in this project. The local government was not allowed to invest in a project that only benefited the retail sector; the public good needed to be served as well.[19] The group of frequent offenders was generally held responsible for a variety of minor offences that affected the general public, like loitering and pickpocketing (Ferwerda, Kleemans, Korf, & Van der Laan, 2003). Connecting the supermarket to a frequent offender database would therefore not only solve a problem in the retail sector, but also contribute to decreasing disturbances caused by a group that was considered a social problem.

As I argue here, this technology served as more than a practical tool. Facial recognition was designed to reorganise the collective of humans and artefacts so as to implement a new logic: 'deterrence'. In the following section I will show that this involved another type of surveillance in which security officers would function as hosts, carry more responsibility for preventing shop crime and monitor a larger target group.

A new logic: Deterrence

The logic of facial recognition was meant to introduce a new mode of operation in the supermarket. Starting in 2001, Dutch police had been trying to implement the deterrence (*tegenhouden*) method.[20] Working according to this concept required 'prevention before the preventive phase' (Raad van Hoofdcommissarissen, 2001). Facial recognition would be an instrument for deterrence because it was not meant to catch people in the act of stealing, but to facilitate a proactive approach to shoplifting. As one of the policy advisors involved in the facial recognition pilot stated:

> 'It isn't some sort of sneaky camera that catches you in the act. That absolutely isn't the goal of this project, absolutely not. It's purely an instrument for deterrence, an instrument that empowers the shopkeeper to actively approach or stop a potential shoplifter. And when you stop it, it hasn't happened, it won't require any negative energy.'

The prevention of crime through proactive behaviour was to be a task of citizens and private organisations. While the paper banning order was an important means to accomplish this in retail crime, the retail sector continued to use its own methods rather than act in accordance with the banning order guidelines.

19. Final draft of the project plan, technology consultant, 2005 (internal document).
20. I follow the Dutch police force in using the term deterrence as the English translation for *tegenhouden* (Board of Chief Commissioners, 2006).

Facial recognition was intended to introduce a new order of surveillance into the supermarket that would redefine the security officer's work practices. First, as one of the city's policy officers put it, facial recognition was a 'sneaky method' to extend the knowledge of shopkeepers beyond the records about offenders they kept themselves. The technology would bring a new target group of police-registered frequent offenders into the supermarket without explicit consent. Consequently, it would direct the gaze of the supermarket guards towards people and types of crime that they regarded as a police responsibility. Second, the security officer would play the part of host to his customers. Third, surveillance would serve not to arrest, but to identify and establish contact with a person. The supermarket's security guards would no longer wait until an item had actually been stolen, and consequently would not have to call in the police. A policy officer described the ideal mode of operation as follows:

> 'When somebody with a banning order walks in, an active strategy of guided shopping should be applied. This is to say you approach the person and request him to leave the shop, and only when this doesn't work, the police can be called in to make an arrest for trespassing … . As for the frequent offenders, if everything works as planned, you will not tell them: "you cannot shop here" or "oh, you're a frequent offender". No, as a security officer you will learn "that's a frequent offender, the police know him, maybe I should get to know that person too", and you will actively remove this person out of the sphere of anonymity by suggesting they use a shopping basket, or by telling them about discounts, walking around with them, or whatever. This will make this person think: "I'm not anonymous anymore, I'm being watched".'

Whereas the logics of collecting strikes and care required the use of CCTV images, knowing the local target group and operating in semi-legal domains, deterrence introduced a new organisational principle that required another type of surveillance. According to this approach, the security officers would no longer stay in the control room, but would be alerted to the target groups in the databases. Consequently, the target group categories of Shoplifters/Normal persons and Shoplifters/Risky individuals would be replaced by the new categories of Shoplifters/Banning orders and Shoplifters/Frequent offenders.

Facial recognition as a reordering device

In this section, I will elaborate on how the technology was used to mobilise the supermarket in the surveillance of a more broadly defined area of petty crime. In the first place, facial recognition offered the opportunity to share police information while complying with privacy regulations. According to the Police Registration Act,[21] sharing police data was only allowed under very strict conditions, including: a contract between public and private actors; timely deletion and updating of information; using the information solely for the purpose of crime control; and an inability of the police to solve the problem without interference of third parties.[22]

With regard to police-registered frequent offenders, the supermarket had already been given fifteen colour photos and names of the most notorious frequent offenders in the neighbourhood known to the police. This was a very laborious method, however, because the paper photos had to be updated regularly and the police had to ensure that the information would not leave the premises of the supermarket. As a matter of fact, at the time of research this method was under debate in light of a court ruling against the use of card games and placemats with frequent offender photos in police canteens.[23]

Meanwhile, the police and the city were still eager to communicate about risk groups with the retail sector. It was thought that facial recognition would solve the problem because it would only display information upon recognition, and the frequent offender database would be updated automatically. Consequently, this technology was regarded as an improvement of privacy. As a policy officer of the city's Department for Community Safety stated:

> '[The frequent offender card game] was eventually forbidden by the *CBP*[24] because a card game is made of paper and will exist for ever... . And what we have now, the interesting thing about it is that we have a database that is not accessible for retailers, they will only be presented with a photo when the frequent offender enters the store and is recognised. They only know what he looks like; they do not know his name, what he has done. They only know: "watch out"... . So this database is also a black box for retailers, they cannot snoop around in it. In comparison with paper, this is an absolute improvement of privacy.'

Moreover, the signing of a contract stating that police information about frequent offenders could be shared on the basis of the Police Registration Act created a new basis for the use of paper posters. The policy staff of the police and the local government argued that if humans would update as rigorously as the computer and if the police would make sure that the paper files are stored in a safe place, there would be no obstacles to the large-scale use of police posters by the retail sector.

---

21. The Police Registration Act (*Wet politieregisters*) was replaced by the Police Data Act (*Wet politiegegevens*) in 2008. This did not affect the pilot discussed in this paper. For an analysis of the new legislation see MacGillavry (2005).
22. Feasibility study of the project, technology consultant, 2005 (internal document).
23. Local newspaper article, published on 9 October 2004.
24. *CBP* is the Dutch acronym for *College bescherming persoonsgegevens*: the Dutch Data Protection Authority.

Furthermore, facial recognition could provide a legal basis for these practices by provoking a court case. During the pilot, the police hoped for a lawsuit on the issue of the production and storage of photographs. To this day, the quality criteria for the use of photos for automatic recognition and their storage in a database remain undefined. As a police participant to the project stated: 'a court ruling would give us all guidelines'. Facial recognition was expected to lead to a court case because it was thought to provoke questions about the trustworthiness of the method, for example, regarding the quality of the police photos and their suitability for automatic recognition. So far, no lawsuit has been filed against this practice or similar applications that make use of facial recognition.
On top of its interaction with privacy regulations, another issue concerning facial recognition was that it required a large number of entries to become effective, even in a pilot situation. In the early phases of the project, however, it was stated that the target group should be defined very strictly. Only 'very active frequent offenders', individuals that had gathered eleven or more police records during the previous five years, including one during the previous twelve months, would be admitted to the database.[25] The supermarket furthermore stated that it would not accept database entries of persons who had been convicted for violent acts because it did not want to expose its security employees to potential harm.
Yet, when the police finally compiled the databases, these arguments were superseded by the argument that the technology 'needs hits', as the head of the project group that defined the target group explained. The chances that the technology would recognise somebody from the database would have been too small if it had only contained the fifteen individuals that the police registered for this area. To increase the chances of recognition, the criteria of admission to the database were relaxed. In the end, the police database included 1,250 multiple, frequent and very frequent offenders from the region.[26] The minimum offence was one case of shoplifting in the last twelve months and two police records in a lifetime.[27] Ultimately, no check for a violent background was included. So in order for the technology to work, it needed to include persons with a minimum of

25. Final draft of the project plan, technology consultant, 2005 (internal document).
26. Various categorisations of frequent offenders existed in Dutch crime policy during my fieldwork period. The categories used by my police informants in this case study were: 1) the first offender: one offence in a lifetime; 2) the multiple offender: two to ten police records in a lifetime, of which one record in the previous twelve months; 3) the frequent offender: more than ten police records in a lifetime, of which one record in the previous 12 months; 4) the very active frequent offender: more than ten police records in the previous five years, of which one record during the previous twelve months.
27. I use the term 'police record' here to refer to the *proces-verbaal*. In Dutch criminal law, the *proces-verbaal* is used as a unit to measure offences. A *proces-verbaal* refers to the police document that reports the nature and details of the offence, the offender and the testimony. This document can report one or more offences. Nonetheless, in the administrative systems of the police the number of *processen-verbaal* is often equaled to the number of criminal records.

two offences in a lifetime, instead of eleven, notwithstanding the nature of the offence.

This suggests that facial recognition should not be considered as a neutral technology that simply links new actors to a network. By evading legal restrictions on the use of paper, this application would enable a reconfiguration of surveillance practices towards a logic of deterrence. In the ideal scenario the security officer would use a database of police information and would assert his presence on the shop floor as a host, thereby employing different monitoring skills than used in the logics of collecting strikes and care. In addition, the technology had its own requirements. It needed hits and so required a larger target group, leading to a quantitative and qualitative redefinition of crime to be addressed by the supermarket. This is how the supermarket became involved in the surveillance of petty crime beyond the store's doorstep. Therefore, the technology is not only a means to further implement the shop ban but it can also be a reordering device for crime control: it was intended to reorganise the relations between actors.

5 A favour that was not asked for

As a reordering device, however, there was resistance to facial recognition. Also, the technology was put to use in unexpected ways. After a year of testing, the city council decided against funding the development of this application any further because it was disappointed with the results of the pilot. During 161 measurement days, four persons were recognised by the technology, so the evaluation report said. In reaction to this result, the local government stated that human recognition appears to be more effective, as the security officers arrested ten individuals with a banning order who were not detected by the technology.[28] The consultants that had supplied the technology disagreed. They claimed that these numbers were being misinterpreted; according to them, the technology itself worked well. In their view, the reason why people with banning orders may not be detected was because the security officers arrested them just before they walk into the camera's view. According to one of the consultants, this happened quite regularly due to the familiarity of the guards with their target group. Some homeless people might even approach the guards in order to be arrested, so that they would have a place to stay for the night, he argued.[29]

The consultants' explanation indicates that this pilot is more than a story about technical failure. Even though this explanation may not fully account for the results of the pilot, they rightly observed that the old logics of surveillance had not been changed by the introduction of facial recognition. This suggests that the use of surveillance technologies does not necessarily lead to the pluralisation of

28. Letter from the mayor to the councillors about the evaluation report, 2008.

29. Press statement of one of the project's corporate consultants, 2008.

policing and surveillance. In this case, the security officers remained focussed on their old target groups: Shoplifters/Normal persons and Shoplifters/Risky individuals. The new databases of Shoplifters/Banning orders and the Shoplifters/Frequent offenders could not replace the advantages of the security officers' contextualised knowledge because they only referred to the type of registration and not to local knowledge and experience. Consequently, these suspect categories did not improve the performance of the logics of collecting strikes and care.

Moreover, the security officers and lower-rank management were simply not convinced that hosting would be an effective strategy against their target group. The guards were furthermore not interested in detecting individuals that were registered by the police for shoplifting and other offences in places other than their own supermarket. This would introduce the problems of the 'outside world' into the supermarket, so they claimed.

Nevertheless, the guards did think that some features of the technology were potentially useful. First, they reckoned that if the system had recognised more individuals, it could have helped to identify one of their target groups: the Shoplifters/Normal persons. The guards argued that 'normal persons' that steal are difficult to recognise, so this is where facial recognition would have been useful. The guards did mention, however, that this would not solve the entire problem, because the technology 'does not recognise thieves', i.e. it cannot observe suspicious behaviour of people that are not part of the database. Second, the guards valued the system's computer as an administrative tool, helping them to store and browse their files.

Higher management at the supermarket head office also valued the technology in a way other than planned. From the beginning, higher management felt uneasy about observing their own customers using facial recognition. They had nevertheless consented to the pilot because they wanted to maintain good relations with the police and local government. Yet in the end the supermarket expressed its interest in using a smart camera system to monitor its goods rather than its customers. As shown earlier in this paper, visibility of products plays an important role in supermarket surveillance. Therefore, the supermarket's security department thought it would be more interesting to know if the technology could be used to alarm the security guards if, for example, a large amount of toothpaste disappears from the racks. After the pilot project had officially ended, the supermarket and the consultant continued to test smart camera systems in the laboratory for the recognition of objects.

6 Conclusion

In this paper, the pluralisation of surveillance and policing was approached through the notion of situated surveillance, in which the observation of certain target groups is understood as the outcome of a heterogeneous collective of objects and humans. I first described the logics of surveillance that were performed by the collectives in which the security officers of the supermarket took part: the logic of collecting strikes and the logic of care. The security officers used various artefacts and types of knowledge to monitor two target groups: the Shoplifters/Normal persons and the Shoplifters/Risky individuals. Moreover, the supermarket security guards carefully kept their distance from police activities.

The facial recognition system examined in this case study was designed to implement an ideal type of crime control, according to the logic of deterrence. By sharing police knowledge, the supermarket would be mobilised to engage in a preventive mode of operation. Facial recognition was thus far from a neutral technology; it was designed to further involve the supermarket in the surveillance of petty crime. It could do so by adapting and evading privacy regulations, in order for the group of Shoplifters/Frequent offenders to be observed by the supermarket's security officers. Moreover, the technology itself required a larger database of offenders, which changed the target groups of the supermarket.

Although facial recognition could potentially intervene in the relations between the various actors involved in monitoring shop crime, the technology could not be implemented in existing practices. This case consequently shows that it is crucial for those that study surveillance to attend to context and agency in order to understand how and why surveillance technologies may be resisted and adapted. The surveillant assemblage rightly draws attention to the diversity and multiplicity of actors that are pulled into networks of surveillance. What is more, Haggerty and Ericson (2000) usefully suggest that technology plays a role in drawing the actors together. However, by assuming a decontextualised logic of surveillance, the exact nature of the processes whereby new actors are drawn into surveillance networks cannot be fully understood. At times, configurations may be far from fluid. Instead, they can be rather viscous and resistant.

As M'charek (2008) argues, the manner in which technologies link actors and articulate suspects depends on the context of usage. This case study highlights several particular ways in which context plays a role. First, the interaction of the surveillants with their target group, 'the community', was an important aspect of the logics of collecting strikes and control. Thus, the surveillant is partly constituted by its subjects. Second, we learn that the mobilisation of new actors is also a process that can include resistance. Third, the notion that technologies never implement ideal strategies in a top-down fashion (Law, 1992) is illustrated by the renewed efforts of the supermarket to use a smart camera system for monitoring goods instead of people.

An important critique of facial recognition is that the technology can contain a built-in preference for the recognition of certain ethnic groups over others. If applied in practice such technological biases can be enforced by the existing discriminatory practices of the surveillants (Introna & Wood, 2004). By focussing on the context of the surveillants' practices, I raise another issue. It should also be recognised that surveillance strategies emerge from coping with a complex situation. In this case, shoplifting is part of a larger problem that includes poverty, homelessness and drug addiction. Policies that implement new regulations and technologies often act on simplified notions of the situation at hand, offering technological fixes that aim to trick new surveillants into assuming more tasks, while creating new grey areas of regulation at the same time. Taking into account the complexity of surveillance practices and how technologies affect these practices may open up new directions for analysis and critique.

References

Akrich, M. (1992). The de-scription of technical objects. In W. E. Bijker, & J. Law (Eds.), *Shaping technology/building society: Studies in sociotechnical change* (pp. 205-224). Cambridge, Massachusetts: MIT Press.

Board of Chief Commissioners (2006). *The police in evolution: Vision of policing*. The Hague. Retrieved April 11, from http://www.politie.nl/ImagesLandelijk/the%20police%20in%20evolution_tcm31-339858.pdf

Boutellier, H. (2002). *De Veiligheidsutopie. Hedendaags onbehagen en verlangen rond misdaad en straf.* Den Haag: Boom Juridische Uitgevers.

Callon, M. (1986a). Some elements of a sociology of translation – Domestication of the scallops and fishermen of St. Brieuc Bay. In J. Law (Ed.), *Power, Action, and Belief – A New Sociology of Knowledge* (pp. 196-233). London: Routledge and Kegan Paul Books.

Callon, M. (1986b). The sociology of an actor-network: The case of the electric vehicle. In M. Callon, J. Law & A. E. Rip (Eds.), *Mapping the dynamics of science and technology* (pp. 19-34) MacMillan.

Deleuze, G. & Guatarri, F. (1987). *A thousand plateaus.* Minneapolis: University of Minnesota Press.

Den Boer, M. (2004). *Out of the blue: Police perspectives on Europe, governance and accountability*. Amsterdam & Apeldoorn: Vrije Universiteit/Politieacademie.

Donoghue, J. (2008). Antisocial behaviour orders (ASBOs) in Britain: Contextualizing risk and reflexive modernization. *Sociology, 42*(2), 337-355.

Dubbeld, L. (2005). The role of technology in shaping CCTV practices. *Information, Communication & Society, 8*(5), 84-100.

Ferwerda, H., Kleemans, E., Korf, D., & Van der Laan, P. (2003). Veelplegers. *Tijdschrift voor Criminologie,* 2(45), 110-118.

Foucault, M. (1995). *Discipline and Punish: The birth of the prison*. New York and Toronto: Random House.

Gad, C., & Lauritsen, P. (2009). Situated surveillance: An ethnographic study of fisheries inspection in Denmark. *Surveillance and Society, 7*(1), 49-57.

Gates, K. (2010). The Tampa "Smart CCTV" experiment. *Culture Unbound, 2*, 67-89.

Grasseni, C. (Ed.). (2007). *Skilled visions: Between apprenticeships and standards*. New York and Oxford: Berghahn Books.

Haggerty, K. D., & Ericson, R. V. (2000). The surveillant assemblage. *British Journal of Sociology, 51*(4), 605-622.

Haggerty, K. D. (2006). Tear down the walls: On demolishing the Panopticon. In D. Lyon (Ed.), *Theorizing surveillance: The Panopticon and beyond* (pp. 46-68). Devon, UK: Willan Publishing.

Haraway, D. (1988). Situated knowledges: The science question in feminism and the privilege of partial perspective. *Feminist Studies, 14*(3), 575-599.

Helten, F., & Fischer, B. (2004). Reactive attention: Video surveillance in Berlin shopping malls. *Surveillance and Society, 2*(2/3), 323-345.

Introna, L. D., & Wood, D. (2004). Picturing algorithmic surveillance: The politics of facial recognition systems. *Surveillance and Society, 2*(2/3), 177-198.

Koops, B.J. (2011), The evolution of privacy law and policy in the Netherlands. *Journal of Comparative Policy Analysis: Research and Practice, 13*(2), 165-179.

Latour, B. (1990). Drawing things together. In M. Lynch, & S. Woolgar (Eds.), *Representation in scientific practice* (pp. 19-68). Cambridge, Mass. and London: MIT Press.

Latour, B., & Hermant, E. (2006). *Paris: Invisible City*. Retrieved November 11, 2010, from www.brunolatour.fr/livres/viii_paris-city-gb.pdf

Law, J. (1992). Notes on the theory of the actor-network: Ordering, strategy, and heterogeneity. *Systems Practice, 5*(4), 378-393.

Loader, I. (2000). Plural policing and democratic governance. *Social and Legal Studies, 9*(3), 323-345.

Lyon, D. (2007). *Surveillance studies: An overview*. Cambridge: Polity.

MacGillavry, E. C. (2005). Heeft u even voor de nieuwe wet politiegegevens? In A. Harteveld, D. H. De Jong & E. Stamhuis (Eds.), *Systeem in ontwikkeling, liber amicorum* (pp. 385-416). Nijmegen: Wolf Legal Publishers.

Marx, G. T. (2002). What's new about the new surveillance? Classifying for change and continuity. *Surveillance and Society, 1*(1), 9-29.

M'charek, A. (2008b). Silent witness, articulate collective: DNA evidence and the inference of visible traits. *Bioethics, 22*(9), 519-528.

Mol, A. (2008). *The logic of care: Health and the problem of patient choice.* London: Routledge.

Mol, A. & Berg, M. (1994). Principles and practices of medicine: The co-existence of various anemias. *Culture, Medicine and Psychiatry*, 18, 247-265.

Norris, C., & Armstrong, G. (1999). *The maximum surveillance society: The rise of CCTV*. Oxford: Berg.

Philips, J.P., Flynn, P.J., Scruggs, T., Bowyer, K.W., Chang, J., Hoffman, K., Marques, J. Min, J., & Worek, W. (2005). Overview of the Face Recognition Grand Challenge. *2005 IEEE Computer Society Conference on Computer Vision and Pattern Recognition (CVPR'05)*,1, 947-954.

Prainsack, B., & Toom, V. (2010). The Prüm regime. Situated Dis/Empowerment in transnational DNA profile exchange. *British Journal of Criminology, 50*, 1117-1135.

Raad van Hoofdcommissarissen (2001). *Misdaad laat zich tegenhouden. Advies over bestrijding en opsporing van criminaliteit.* Retrieved April 11, 2011, from http://www.politie.nl/ImagesLandelijk/misdaad%20laat%20zich%20tegenhouden_tcm31-66184.pdf

Rose, N. (2000). Government and control. *British Journal of Criminology, 40*, 321-339.

Schuilenburg, M., & Van Calster, P. (2009). De collectieve winkelontzegging: Een antwoord an willekeur op overlast. In H. Boutellier (Ed.), *Omstreden ruimte. Over de organisatie van spontaniteit en veiligheid* (pp. 137-155). Amsterdam: Van Gennep.

Smith, J. D. G. (2004). Behind the screens: Examining constructions of deviance and informal practices among CCTV control room operators in the UK. *Surveillance and Society, 2*(2/3), 376-395.

Van Steden, R. (2007). *Privatizing policing: Describing and explaining the growth of private security*. Den Haag: Boom Juridische Uitgevers.

Van Stokkom, B. (2009). Verwijderd van de straat. Over veiligheid door gebiedsverboden. In H. Boutellier (Ed.), *Omstreden ruimte. Over de organisatie van spontaniteit en veiligheid* (pp. 99-118). Amsterdam: Van Gennep.

'Better Safe than Sorry, But You Know, Don't Overdo It'. Responsibilisation and Fatalism in Perception of Safety

*Evelien Van den Herrewegen**

1 Introduction

The slogan 'Better Safe than Sorry, But You Know, Don't Overdo It'[1] is inspired by a column written by Melissa Summers about 'National Baby Safety Month' (2007). In a cynical but funny manner Summers criticises the industry that, in order to sell some outlandish safety devices,[2] preys on one of parents' biggest fears: a preventable accident hurting their baby. Summers does not ridicule their fear, but urges parents not to get terrorised by these messages, and instead to set out some practical ground rules to protect infants. Although this article is not focused on 'baby safety', I will show that people use the same slogan in dealing with crime. While acknowledging that people are (more) aware of crime, and take precautions, I will note that this awareness and this behaviour does not necessarily have a dramatic impact on people's daily lives, nor are they to be interpreted as signals for a problematic perception of safety. Parallel to the conclusions of an Australian study by Deborah Lupton (2000), my own research findings in a Flemish neighbourhood raise some strong reservations to the current theoretical assumption that people nowadays are caught up in a 'Culture of Fear' (Furedi, 2006) or victims of a 'Culture of Control' (Garland, 2001), and that 'fear of crime' functions as a way to project (Hollway & Jefferson, 1997) or to express (Jackson, 2004) late modern ambiguous anxieties. In the conclusion I will discuss the relevance of this interpretative research for surveillance studies.

* Evelien Van den Herrewegen is post-doctoral researcher at the University of Antwerp, lecturer at Courtrai University College, and member of the Research Unit Social Analysis of Security (Ghent University).

1. The original title is: 'Be Safe, But You Know, Don't Overdo It.': http://alphamom.com/parenting/parenting-health-safety/be-safe-but-you-know-dont-overdo-it-national-baby-safety-month/ [11/03/2011]
2. One of these devices is the Thudguard, an infant safety hat.

2 A Cultural Preoccupation with 'Being Safe'

In his book 'Culture of Fear' (2006), the British sociologist Frank Furedi argues that we need to understand the current western preoccupation with 'being safe', and its resulting fears (such as the parental fear for their baby's safety) not as reactions to clear and present dangers such as diseases, unemployment, disaster or criminality. This preoccupation and these fears are in fact a result of a new culture characterised by a mistrust in humanity and its ability to plan and change the future, i.e. the decline of social engineering. In contrast to past generations and non-western cultures, people in the western industrialised world have a declining belief in shared and general frameworks to understand and deal with threatening situations: 'Our reaction has become so private and personal as we are lacking great narratives to explain the dangers we are exposed to.' (Furedi, 2006, p. 38, own translation). Given this lack of consensus and general knowledge to detect and deal with risks, people begin to question every aspect of daily life and as such even the most simple things of life such as eating or sunbathing, are turned into hazardous activities. Moreover, this new culture, characterised by uncertainty and high risk awareness, is enhanced by a policy and a morality that urge people to be careful and display precautionary behaviour, and that condemn them when they expose themselves or others to danger: '"Better to be safe than sorry" is the motto nowadays. And as this kind of preoccupation with safety has become routine and perfectly normal, then nothing is safe anymore' (Ibid, p. 45, own translation).

In his book, *'A Culture of Control'* (2001), David Garland makes a similar point, although his analysis is more focused on the criminological implications of the western preoccupation with 'being safe', i.e. the transformations within the field of crime control. Although Garland describes and recognises that this public awareness and these transformations are influenced and enhanced by a number of factors and actors (media, politicians) and processes (neoliberal thinking, privatisation), he also argues, like Furedi, that these factors and processes could only flourish because of a new cultural practice in dealing with criminal risks and victimisation. Starting in the 1970s, but thriving in the 1980s, crime became a prominent feature in people's daily life, and this aspect has particularly changed the middle-class's attitudes towards crime and crime control:

> '... fear of crime became an established part of daily existence. What were once, for much of the middle-class population, fleeting, occasional fears, linked to particular situations and unusual circumstances, now became much more routine, much more part of the habitus of everyday life, particularly in large cities.' (Garland, 2001, pp. 152-153).

This heightened 'crime-consciousness' is hard to change as it 'comes to be embedded in everyday social life and institutionalised in the media, in popular culture and in the built environment.' (Garland & Sparks, 2000, p. 16). As such, contemporary society is characterised by a 'crime complex', that has detrimental effects on people's lives and feelings:

> 'Citizens became crime-conscious, attuned to the crime problem, and many exhibit high levels of fear and anxiety. They are caught up in institutions and daily practices that require them to take on the identity of (actual or potential) crime victims, and to think, feel and act accordingly.' (Garland, 2001, p. 164).

According to Garland (2001), this heightened preoccupation with crime produces two new contrasting reactions. On the one hand a *'criminology of everyday life'* that can be described as late modern as it de-dramatises crime, treating it as a *predictable* outcome of normal routines of daily life, and as such urges institutions as well as citizens to *adapt* to these new circumstances. Consequently, safety is everybody's concern and duty, i.e. it is people's responsibility to avoid unnecessary risks and take precautions to protect themselves and (significant) others. On the other hand, Garland distinguishes a *'criminology of the other'*, an anti-modern reaction as it re-dramatises crime by depicting criminal actions as voluntary, bad choices of wicked individuals, and as such it calls for tougher policing and harsher sentencing to protect the 'innocent' public from the dangerous 'others'. In this sense, crime control becomes a form of *acting* out to reassure the worried and outraged public. Although the reactions differ substantially, both confirm and enhance the cultural preoccupation with crime. The aim of this research is not to examine people's punitive reactions towards criminals, but people's perceptions of safety and how they deal with crime in their daily lives instead. Therefore, the focus is on Garland's 'criminology of everyday life', and to examine empirically the normalisation of crime in people's daily lives. But first we turn to the 'fear of crime' tradition that assumes that this new cultural preoccupation with safety has made people more sensitive about crime and leads to a problematic perception of safety.

3 'Fear of Crime' as a Late Modern Concept

According to Furedi and Garland, people's preoccupation with safety is not so much a reaction to specific threatening experiences, but needs to be understood as a new cultural attitude and practice towards safety, and crime in particular. This theoretical assumption has recently also entered the domain of 'fear of crime' research. Since its 'discovery' in 1965 (Lee, 2007), this research tradition has been struggling with a strange paradox, i.e. the observation that people with the lowest risk of criminal victimisation exhibit the highest fear (e.g. women and elderly), whereas people with a higher risk of victimisation have less fear (e.g. young men) (Vanderveen, 2006). In short, people's feelings (fear) do not always correspondent with their experience of crime. In order to answer this 'fear of crime paradox', researchers broadened their scope and included non-criminal factors to explain this fear. Synthesising almost 50 years of etiological 'fear of crime' research, these factors can be roughly organised in four broad models to explain 'fear of crime'.[3] The *'victimisation'* model focuses on the influence of direct but also indirect experiences of crime (e.g. victimisation of friends, relatives, media coverage). The *'vulnerability'* model states that certain individuals are physically (e.g. gender, age) and/or socially (e.g. socio-economic status, ethnic origin) more vulnerable to crime and thus to fear of crime. The *'environmental'* model examines the physical (e.g. disorder) and/or social organisation (e.g. lack of informal social control) of one's neighbourhood and its effects on people's perceptions of safety. However, it is only recently, in what I will call the *'uncertainty'* model, that 'fear of crime' is not solely considered as a direct emotional reaction to crime, or other deviant behaviour, and a lack of sources to cope with these experiences, but as a manifestation of a broader sense of non-well-being inherent to late modern society.

Rooted in sociological theories developed by Ulrich Beck (1992), Anthony Giddens (1990), and Zygmunt Bauman (2000), the assumption is that 'fear of crime' is a result of a late modern anxiety that is triggered by a gap between people's heightened awareness of risks and their uncertainty about how to handle these risks. Bauman (2000) in particular states that people's incapacity to control the future (uncertainty), and their lack of resources to deal with risks (insecurity) are channelled into concerns about safety, of one's body, family and property. As the roots of uncertainty and insecurity are left intact, this preoccupation with safety will not ease people's mind. Because tools to tackle this uncertainty and insecurity are lacking, citizens as well as policymakers seem to focus their attention on sources of fear that are identifiable and assignable, i.e. the criminal 'other'. In this sense, it is argued that 'fear of crime' has become 'the Leitmotiv of a society that does not tolerate any form of risk and uncertainty' (Pleysier, 2009, p. 173, own translation).

In 'fear of crime' research, Hollway & Jefferson (1997) were one of the first to examine the relationship between late modern anxiety and people's fear of crime. Based on a qualitative study they argue, parallel to Bauman's thesis, that the preoccupation with crime is a way to cope with late modern anxieties. Unlike late modern risks, crime and crime-related issues function as a relative familiar domain with identifiable victims and blameable culprits that are manageable and potentially controllable. As such, Hollway & Jefferson conclude that crime 'serve[s] unconsciously as a relatively reassuring site for displaced anxieties

3. We will briefly discuss these four models, but for a more thorough review we refer to Hale (1996), Ditton & Farrall (2000) but also Vanderveen (2006) and Pleysier (2009).

which otherwise would be too threatening to cope with' (Hollway & Jefferson, 1997, p. 264). 'Fear of crime' is thus seen as a projection of a formless and ambiguous feeling of uneasiness and uncertainty about one's position and identity in society.

Recently, quantitative researchers into fear of crime have also been integrating these late modern theories to indicate that the prevalence of fear of crime has been exaggerated (Farrall & Gadd, 2004). Several studies show that 'classic' survey questions[4] not only intercept the direct emotional reaction to situated and temporary experiences (e.g criminal victimisation), but also tap into a more stable and general predisposition that is an expression of people's broader social concerns, values and attitudes towards a rapidly changing society (Gray, Jackson, & Farrall, 2008). Following developments in psychology (Gabriel & Greve, 2003), new measurement tools have been devised that make a distinction between this 'narrowed' or 'experienced' fear, and a 'broader' or 'expressive' fear (see also Pleysier, 2009). As such, it is argued that we need to 'treat fear as an expression of related concerns, funnelled through this concept of crime (expressive fear), as much as summed expression of threat and vulnerability (experienced fear)' (Jackson, 2004, pp.962-963).

In sum, the assumption is again that people's perceptions of safety (fears, concerns, worries) need to be explained as an expression of other diffuse anxieties that are triggered by threats to our certainty (e.g. financial and economic crises) and security (e.g. breakdown of the welfare state). Although I do not question these late modern transformations, nor deny that dealing with safety and crime in particular has become a prominent aspect in contemporary society, I do want to examine more closely whether this cultural preoccupation with 'being safe' has detrimental effects on people's daily lives and results in a problematic perception of safety.

4 People's Perceptions of Safety in Everyday Life

Our own study, although not an exact replicate, is very much drawn from an Australian research conducted by Deborah Lupton (2000). In this study, that was part of a wider study on fear of crime, Lupton examines the relationship between people's risk perceptions of being a victim of crime and their worries about being a victim of these crimes. Based on a quantitative survey with 148 respondents, Lupton confirms the findings of Ferraro & LaGrange (1987) that people's risk perception is not necessarily a surrogate for fear, i.e. 'Individuals may feel that they are highly likely to be a victim of a particular crime, but may not be particu-

4. E.g.: 'How safe do you feel walking in your neighbourhood at night?'

larly worried or fearful about the possibility, and vice versa.' (Lupton, 2000, p. 23). However, the more interesting results stem from in-depth interviews with 65 of these respondents. In these interviews respondents were asked to explain their views and elaborate on how they assess their personal risk and their worries about crime in the contexts of their everyday lives. Two important but seemingly paradoxical accounts about criminal victimisation were distinguished:

> 'They [respondents] moved between the notion that victims may in some way [*sic* be] responsible for the crime, by not engaging in proper defensive strategies or having the 'wrong attitude', and the fatalistic notion that crime is random and unpredictable, and could happen to anyone, regardless of the efforts they took to prevent it' (Lupton, 2000, p. 30).

In other words, Lupton's respondents indicate that on the one hand crime is indeed a predictable and thus controllable aspect of daily life. Consequently, people believe that they are responsible for taking precautionary measures, such as locking doors, and adopt proper behaviour, such as not drawing attention to yourself. On the other hand, however, people also use fatalistic notions when describing their chances of being a victim of crime. As such, people also accept that, despite their precautionary behaviour, crime will happen anyway as it is so common (e.g. car theft) that it is just a matter of time that one is victimised, or so rare (e.g. murder) that is just a matter of bad luck or fate. By using these fatalistic notions people deliberately try to avoid worrying too much about crime and the risk of criminal victimisation. Hence, Lupton argues, the often-found 'difference between perceived risk and levels of worry about the risk' (Ibid, p. 31). Although these accounts may seem at odds, Lupton argues that this pre-modern discourse of fatalism enables people to actively cope with the late modern context in which a heightened crime awareness and the responsibility for self-reliance is stimulated. In sum, people use fatalistic notions to prevent becoming victims of a 'Culture of Fear' or a 'Culture of Control'.

5 Research Design

The aim of my own research is to investigate whether I can confirm Lupton's findings and conclusions in the context of a social housing estate in a provincial and affluent town in Flanders (Dutch speaking region in Belgium – Europe). In this 35-year-old estate there are 334 houses: 46.4% of the homes are rentals, 49,1% are owned and 4.5% are private houses, but on plots purchased at a reduced price. There are about 800 residents, and compared to the town average, the estate has seen a sharp rise in the ageing population (31% versus 50% of the residents are aged 55 or older). Relatively few residents have a non-Flemish or non-Belgian background. The estate is not a high crime area and generally described as a peaceful and pleasant neighbourhood. The main concerns are traffic safety (speeding), street litter, and loitering youth who, at least according to some residents, are also involved in drug use and drug dealing.

The research is based on in-depth interviews with 27 residents of this housing estate. On average interviews took almost two hours. The proportion of men and women interviewed was roughly equal (14 were female, and 13 were men). The average age was 47, with 77 being the oldest respondent and 21 the youngest. Most respondents are homeowners (20), and seven respondents are social tenants. Regarding family composition, most of the respondents were married (17), four of them were divorced mothers with children in their teens, three respondents were in their twenties but still live with their parents, two women were widows and one respondent was a 40 year old bachelor who has never been married before. Professionally, most respondents were employed in low-level white-collar jobs (21); only three respondents were blue-collar workers. However, almost half of the respondents were currently not active on the labour market because of retirement (10) or sick leave (2), or because they are studying (2) or doing unpaid work in the house (1).

At the start of the interviews, respondents were asked what they associate with 'safety'. Although respondents usually talked about crime and crime-related issues such as disorder, some of them also spontaneously referred to traffic safety. Next, the respondents were invited to elaborate on what they had indicated as unsafe situations, places or persons. In particular, respondents were requested to account for their associations with safety, and to explain how they coped with unsafe situations or prevent being unsafe. Finally, respondents were also invited to talk about change in their perception of safety, i.e. how this perception had changed during the course of their life, or how it probably would change in the future.

In this research an interpretative framework was used in which perception of safety was considered as a process of interactive interpretative practices. In this sense, perception of safety is not studied as an outcome of factors, people or processes independent of human interpretation and action. We thus distance ourselves from the traditional quantitative approach that conceptualises perception of safety as a reaction to a complex interplay of individual characteristics (e.g. age and gender), environmental cues (e.g. disorder and lack of informal control) or political and socio-economic processes (e.g. neo-liberalism). Conversely, Verfaillie and I (2011) claim that, to understand people's perceptions of safety, it is necessary to place these perceptions within the context of people's everyday lives, and to study how people construct their perception of safety in interaction with others. This entails that we not merely map out 'what' respondents describe as unsafe (i.e. dark alleys, public transport stations, strangers, being alone at home, burglary, assault) but try to unravel 'how' people give meaning to unsafe experiences. In other words, the analysis is focused on how the described unsafe situation deviates from people's normal routines and expectations, and how in dealing with their perception of safety people use and believe in certain strategies.

Based on this research design, the aim is to examine the assumptions in sociological and criminological theories that people are more conscious of crime risk, and more susceptible to the appeal of individual responsibilisation. Following the assumptions of recent 'fear of crime' research, the central question is whether this engrossment with crime and individual responsibility has detrimental effects on people's daily lives, and manifest themselves as a problematic perception of safety.

6 The Vigilant and Responsible Citizen: 'Better Safe than Sorry'

Respondents' accounts of their perception of safety unequivocally show that there is indeed a higher crime consciousness: either because they identify a 'real' rise in crime, or because respondents are convinced that due to certain factors, often they refer to the media, people are more informed about crime incidents:

> 'It's not that it [safety] was so much better in the past. Then people were hardly informed and kept in the dark. But, I don't think it's so much better now: ... it's all about sensationalism these days.' (Marc, 64-year-old retired college teacher).

Consequently, many respondents point out that crime has become part of daily life, i.e. being a victim of crime is something that can happen to anyone anywhere:

> 'I know that Brussels [capital of Belgium] is not that safe [respondent tells that he was a victim of a car break-in]. Yes, I think nowadays everybody has experienced something similar.' (Rik, 58-year-old public servant).

Alongside this normalisation of crime, respondents also have a strong sense of individual responsibilisation to prevent becoming a victim of crime. As such, people quite easily adopt precautionary attitudes and behaviour in their everyday life.

> 'These days you can't keep your car unlocked. I think there's a good chance that in the morning your car radio is stolen, or the entire car is gone.' (Guido, a 63-year-old retired municipal worker).

Some respondents even claim that safeguarding is primarily their own responsibility, and not that of the police or other official authorities:

> 'Being safe is my own responsibility. … Who else will make sure I'll be safe? A police officer? … No, no, it's up to you to make sure you're safe.' (John, a 77-year-old retired office clerk).

Within this normalisation and responsibilisation discourse, respondents describe unsafe situations as identifiable, and criminal victimisation as avoidable. 'Better Safe than Sorry' is thus an often-used expression among the respondents. The self-evident manner in which respondents talk about crime and precautionary behaviour is a strong indication for the social embeddedness of the 'vigilant and responsible citizen', i.e. a citizen that acknowledges the fact that 'the world has become a dangerous place' (John) and automatically adapts to it:

> 'I think that break-ins are on the rise, so locking doors becomes a matter of course. I think it's a logical consequence of what we have been experiencing.' (Guido).

In this sense the respondents' accounts confirm the 'criminology of everyday life' identified by Garland, i.e. crime is accepted as a normal fact of daily life that people can avoid by acting responsibly. Our findings thus confirm the existence of the 'vigilant and responsible citizen'. However, although respondents admit having a high or increased crime awareness and describe several precautionary measures, they also indicate that these thoughts and this behaviour do not necessarily have a negative impact on their daily life: 'I do not have the impression that I live differently or that I feel unsafe' (Rik). Moreover, respondents do not automatically link this 'criminology of everyday life' with a problematic perception of safety, i.e. feeling constantly fearful of, or worried about, criminal victimisation:

> 'I certainly do not feel unsafe. … No, I am certainly not more fixated on it [safety]. We [he and his wife] live our lives totally normally. If we want to go out, we go out. You just shut the front door and that's it. Unsafe, no. Certainly not. No, no.' (Guido).

As such respondents indicate that being careful and taking precautions keep them from becoming too fearful or too worried, and some of them even state that taking precautionary measures is enhancing their quality of life. In this sense, we notice that 'the vigilant and responsible citizen' actively seeks out strategies that are not detrimental and a function of his/her everyday life. In recent quantitative studies, Jackson et al. (Gray, Jackson, & Farrall, 2010; Jackson & Gray, 2010) make a similar observation and recommend making a distinction between functional and dysfunctional worry about crime:

> '"Dysfunctional worry" occurred when participants took precautions to protect themselves from crime, but reported that these actions reduced their quality of life. By contrast, "functional worry" resulted in them taking precautions, feeling safer, but without their quality of life being reduced by the precautions they took or the levels of worry they experienced in relation to this.' (Gray, Jackson & Farrall, 2010, p. 4).

In sum, these studies confirm that concern about crime does not automatically have a negative effect on people's lives, but, conversely, can also be a positive force that urges people to be vigilant and responsible. However, this raises the question of how to make a distinction between healthy and unhealthy self-protection attitudes and practices, and moreover, how to differentiate between a problematic and non-problematic perception of safety.

Returning to my own empirical findings, I observed that respondents admit that, although they currently have a non-problematic perception of safety, criminal victimisation would turn their concerns about crime from a positive force into a damaging experience: 'I think that experiencing something dreadful might really turn your life upside down' (Martine). By the same token, Jackson and his colleagues find some early evidence that 'recent experience of crime helps "tip" worry about crime from something that is motivational (functional) into something that erodes well-being and quality of life (dysfunctional)' (Jackson & Gray, 2010, p. 15). Although I do not deny that criminal victimisation can be a devastating experience, my own research findings indicate that not every victimisation is perceived as a traumatic event and linked with a problematic perception of safety.

As an example I focus on Emma, a 43-year-old woman who after withdrawing money from a cash dispenser became a victim of a violent hold-up. Although she lost the money, and ended up with bruises and a fractured jawbone, Emma talks rather lightly about the incident, even laughing a couple of times. In her account she clearly indicates how she could have prevented victimisation. First, she tells how 'stupid' she was to be withdrawing money on a Saturday night. Then she indicates that she should have been more attentive and should have noticed that three men were watching her. Finally, she admits that refusing to give the money and defending herself was not such a good idea as 'three against one was a bit too much'. Throughout her account we notice that Emma knows very well how to prevent being a victim of a hold-up, and how to react to it if such an event occurs. Emma thus knows very well how to be a 'vigilant and responsible citizen', and consequently blames herself for her victimisation. However, this self-reproach has no negative impact on her life or her perception of safety as the precautionary attitudes and behaviour, typical for the 'vigilant and responsible citizen', are not damaged by the victimisation. On the contrary, these attitudes and behaviour are confirmed. In this sense, I argue that mere victimisation is not sufficient to turn a non-problematic perception of safety into a problematic one.

The account of Heidi, a 28-year-old woman, demonstrates a problematic perception of safety.

Heidi is very worried about burglary and considers the consequences of such an event as very serious, i.e. not be able to go to work, move house immediately. Although she has not been a victim herself, she refers to burglary incidents experienced by relatives and co-workers. Heidi would like to install roll-down shutters, but at the present this is financially not feasible, so she comforts herself by locking doors, mapping out escape routes, and she also refers to the informal social control exercised by her neighbours. As such, Heidi differs little from other respondents that worry about burglary and develop all kinds of strategies to protect their house. But contrary to the other respondents, these strategies do not ease Heidi's mind: she sleeps very lightly, and when her husband goes to work in the evening, her sister has to keep her company. Heidi is very annoyed about this, particularly the fact that she, as a grown woman, cannot be home alone late at night. Consequently, I think that Heidi has a problematic perception of safety. Not because she has been victimised, or because she has no strategies, but because she is not convinced that the precautionary measures will protect her from becoming a victim of burglary. In this sense, I suggest that to identify a problematic perception of safety, we should not merely study people's access to strategies (e.g. having the financial means to install a burglary alarm), but also map out the extent people believe that there are strategies that will protect them from becoming a victim of crime.

Carvalho & Lewis (2003) have a similar notion of a problematic perception of safety. Based on a qualitative study about the 'fear of crime' among Afro-American women living in impoverished urban neighbourhoods in Chicago, they observed that all these women are aware of crime and disorder in their neighbourhoods and take precautionary measures. However, only a few of them admitted being very fearful and all of these fearful women showed a large degree of uncertainty about the effectiveness of their safety measures. In sum, Carvalho & Lewis (2003, p. 807) conclude that 'not only the uncertainties of a situation (its extraordinary and random nature), but also the uncertainties of how to address it properly can contribute to fear'. However, we must not confuse this lack of confidence in safety strategies, with another strategy that people sometimes use to keep their perception of safety in their daily lives functional – one that might even help Heidi to cope with her fear of burglary – i.e. employing the pre-modern discourse of fatalism.

7 The Fatalistic Citizen: 'But You Know, Don't Overdo It: You Can't Prevent and Control Everything!'

The late modern discourse and its image of the 'vigilant and responsible' citizen entails that crime is preventable and that people need to uphold the motto 'Better Safe than Sorry'. However, although respondents strongly believe in this discourse and its accompanying precautionary attitude, they also indicate that vigilantism and acting responsibly are not always effective as crime is not always predictable and often a matter of bad luck and fate. Like Lupton's research, our findings show that respondents refer to very common crimes such as burglary and theft:

> 'It all makes little difference. If they want to break-in, they will find a way to come in. It's part of living in our society, and we'd better get used to it.' (Rik, while talking about his experience as a burglary victim).

But equally, respondents refer to exceptionally situations, in particular they talk about extreme public events, such as 9/11, public shooting massacres or acts of meaningless violence:

> 'Those acts are committed by people who are not normal. And to what extent is that avoidable? What can we do about it? There isn't a lot that we can do, is there? (Alice, 54-year-old widow).

These findings, parallel to those of Lupton's, question the assumption that crime functions as an unconscious coping mechanism for late modern anxieties because crime risks are predictable and potentially controllable. According to Lupton's Australian respondents as well as the Flemish respondents in this research, some risks are indeed calculable and therefore actionable, but some of these risks are not and are just a matter of fate and bad luck. In this sense, the late modern discourse of normalisation and responsibilisation is combined with a pre-modern discourse of fatalism. Although these discourses might at first seem at odds, Lupton argues that this is not the case if we take into account that the late modern discourse not only intensifies people's (crime) risk awareness and desire to control, but also describes individuals as 'active rather than passive, as in control of one's life rather than under the sway of others' (Lupton, 2000, p. 33). As 'vigilant and responsible citizens', respondents thus actively seek out strategies to prevent becoming a victim of crime or suffering from fear of it. However, these strategies are not limited to what is conventionally seen as proper and effective crime control strategies, i.e. avoiding high crime areas, installing security locks and so on. Realising that some crimes are so common, or so random, respondents take on a resigned attitude towards crime to avoid becoming too engrossed by crime and crime control, and as such they can get on with their daily lives. This is wel illustrated by the account of a 22-year-old female respondent who talked about a recent shooting incident in which a young man (Kim De Gelder) entered a day care centre, killed a child care worker and two babies aged 9 and 6 months, and injured twelve others (ten children and two adults). Virginie is still shocked about the incident, but she also tries to put things into perspective:

> 'In the long run, you have to let go, ... because otherwise you'd be anxious everywhere. ... and you'd have to lock yourself indoors all the time. ... But I think that if you continually worry about these things, you simply become paranoid.' (Virginie, 22-year-old woman).

Fatalism thus enables people to handle their high crime awareness and the pressure to act responsibly, i.e. by resigning to 'bad luck' and 'fate' people prevent the late modern discourse of normalisation and responsibilisation from dominating their daily lives, and becoming an unhealthy perception of safety. As such the motto 'Better Safe than Sorry' is generally complemented with the guideline: 'But, You Know, Don't Overdo It'. This means that 'the vigilant and responsible citizen', after taking the necessary measures to prevent victimisation, accepts his/her fate as 'You Can't Prevent and Control Everything'.

Respondents tended to mention this pre-modern fatalistic discourse quite often. In this sense Summer's adding of 'You Know' to the guideline 'But, You Know, Don't Overdo It' is very significant. By using 'You Know', Summers appeals to the commonsense of parents, i.e. even though it is normal and parents are expected to be concerned about, and sometimes fearful for their children's safety, parents also *know* that it is not acceptable to exaggerate those fears and act upon every single one of them. By the same token, in dealing with crime, the use of fatalism is as much a socially and culturally shared principle as the need to act responsibly. Therefore, these findings indicate that responsibilisation and fatalism are both cultural practices that people use to prevent becoming fearful creatures. While the late modern discourse of responsibilisation encourages people to believe in their own ability to develop and adopt attitudes and precautionary measures to prevent victimisation, the pre-modern discourse of fatalism helps them to ease this pressure, and thus 'cope with the burden of fear and personal responsibility for self-protection' (Lupton, 2000, p. 35).

8 Conclusion

The aim of this paper was to examine how people deal with crime in their everyday lives. Inspired by the work of Deborah Lupton, and based on own research findings, the conclusion is that people are not necessarily 'cultural dopes' (Garfinkel, 1967) that are caught up in a 'Culture of Fear' or a 'Culture of Control' that demands them to think and act as (potential) crime victims and that induces a problematic perception of safety. Although people are indeed more crime conscious and susceptible to the appeal of responsibilisation, they do not automatically experience it as an assault on their everyday life. As 'vigilant and responsible citizens', people seek out strategies to predict, to avoid and even to control criminal victimisation. But as active agents, people also develop strategies, i.e. fatalistic notions, to prevent that this preoccupation with 'being safe' from having a detrimental effect on their daily lives. People thus deploy late modern as well as pre-modern strategies to avoid becoming fearful creatures.

Contrary to current theories and studies into 'fear of crime', this research has demonstrated that a problematic perception of safety cannot be simply reduced to a projection or an expression of ambiguous anxieties and uncertainties about a changing society. Nor can it be put down to a lack of strategies to deal with these changes. People are not passive victims of their own characteristics (e.g. age, gender, and socio-economic status) and living environment, nor are they passively subjected to social, economic and political transformations. As such, we need an interpretative research programme that acknowledges the complex and human character of perceptions, and does not reduce them to a concept influenced by factors and processes outside human interpretation. Instead it should take these perceptions seriously and address them. This entails respecting individual agency and viewing people's perceptions from the context of their daily lives and examining how people manage these perceptions. However, this does not mean that we can only make statements about individual perceptions of safety. The interpretative programme also acknowledges the fact that people's perceptions are not formed in a social vacuum, but are developed in interaction with others and thus based on socially and culturally shared frameworks. As such, although always temporary and occurring within a certain setting (e.g. a neighbourhood or a group), we can unravel the shared notions on which perceptions are based, and thus make more general claims about people's perceptions and develop specific measures to help people deal with them.

Although this research did not focus on how people deal with surveillance techniques, I would like to end with some important suggestions for surveillance studies. In this research domain, too, the question may be raised of whether surveillance techniques have a positive or negative effect on people's perceptions of safety. To understand the relationship between surveillance and people's perceptions, we need to focus on how people, within a particular location and time frame, perceive the crimes that are under surveillance and examine how they make use of these surveillance techniques in dealing with these crimes. If, for example, we want to study how CCTV has an effect on people's perception of pick pocketing or loitering youths in a shopping mall, we first need to examine how they perceive these incidents and places: to what extent do these deviate from people's normal routines and expectations (e.g. is there a difference between first time visitors or frequent mall users and shop owners?). Also, how do these shopping mall users deal with these incidents and do they believe in the available strategies to avoid becoming a victim (e.g. keeping a close eye on your bag, avoiding certain places etc.). Next, the study needs to map out how the surveillance technique is part of these routines and strategies, i.e. to what extent do people believe that CCTV is a useful mechanism for protecting them from becoming victims and how does it relate to other strategies? Or conversely: do they think CCTV indicates or confirms that the shopping mall is unsafe and to be avoided?

The results of such studies might be surprising. Often it is claimed that a society obsessed with crime and safety is a society infatuated with surveillance and has a great belief (and financial investment) in surveillance techniques. However, intentionally or unintentionally, surveillance techniques can also be detrimental to people's perceptions of safety, as they stimulate suspicion and distrust because every person is a potential culprit and every social interaction is to be regarded as a potential hazard (Prior, 2005). Bearing in mind the findings in this research and that of Lupton, it may turn out that, although people may appreciate additional techniques that supplement their human senses and help them to take precautions, they may also prefer to use fatalistic notions to deal with crime risks instead of turning their houses and living environments into unliveable and asocial fortresses.

References

Bauman, Z. (2000). *Liquid Modernity*. Cambridge: Polity Press.

Beck, U. (1992). *Risk Society: Towards a New Modernity*. London: Sage.

Carvalho, I., & Lewis, D. A. (2003). Beyond Community: Reactions to Crime and Disorder Among Inner-city Residents. *Criminology, 41*(3), 779-812.

Ditton, J., & Farrall, S. (2000). *The fear on crime*. Dartmouth: Ashgate.

Farrall, S., & Gadd, D. (2004). Research Note. The Frequency of the Fear of Crime. *British Journal of Criminology, 44*, 127-132.

Ferraro, K. F., & LaGrange, R. L. (1987). The Measurement of Fear of Crime. *Sociological Inquiry, 57*, 70-101.

Furedi, F. (2006). *Cultuur van angst* (N. Groen, A. Mulder & M. Post, Trans.). Amsterdam: Meulenhoff.

Gabriel, U., & Greve, W. (2003). The Psychology of Fear of Crime. *British Journal of Criminology, 43*(600-614).

Garfinkel, H. (1967). *Studies in ethnomethodology*. Cambridge: Blackwell Publishing Ltd.

Garland, D. (2001). *The Culture of Control. Crime and Social Order in Contemporary Society*. Oxford: Oxford University Press.

Giddens, A. (1990). *The Consequences of Modernity*. Cambridge: Polity Press.

Gray, E., Jackson, J., & Farrall, S. (2008). Reassessing the Fear of Crime. *European Journal of Criminology, 5*(3), 363-380.

Gray, E., Jackson, J., & Farrall, S. (2010). Feelings and Functions in the Fear of Crime. *British Journal of Criminology, Advance Access published* 1-20.

Hale, C. (1996). Fear of Crime: a Review of the Literature. *International Review of Victimology, 4*, 79-150.

Hollway, W., & Jefferson, T. (1997). The Risk Society in an Age of Anxiety: Situating Fear of Crime. *British Journal of Sociology, 48*(2), 255-266.

Jackson, J. (2004). Experience and Expression. Social and Cultural Significance in the Fear of Crime. *British Journal of Criminology, 44*, 946.

Jackson, J., & Gray, E. (2010). Functional Fear and Public Insecurities about Crime. *British Journal of Criminology, 50*, 1-22.

Lee, M. (2007). *Inventing Fear of Crime. Criminology and the Politics of Anxiety*. Devon: Willan Publishing.

Lupton, D. (2000). Part of Living in the Late Twentieth Century: Notions of Risk and Fear in Relation to Crime. *Australian and New Zealand Journal of Criminology, 33*(1), 21-36.

Pleysier, S. (2009). *'Angst voor criminaliteit' onderzocht. De brede schemerzone tussen alledaagse realiteit en irrationeel fantoom*. Proefschrift tot het verkrijgen van de graad van doctor in de Criminologische Wetenschappen.(promotor: prof.dr. G. Vervaeke; co-promotor: prof.dr. J. Goethals): Afdeling Strafrecht, Strafvordering en Criminologie – Faculteit Rechtsgeleerdheid – Katholieke Universiteit Leuven.

Prior, D. (2005). Civil Renewal and Community Safety: Virtuous Policy Spiral or Dynamic of Exclusion? *Social Policy & Society, 4*(4), 357.

Van den Herrewegen, E., & Verfaillie, K. (2011). Over waarheidsaanspraken in het veiligheidsdomein. De kloof tussen objectieve en subjectieve onveiligheid anders bekeken *Panopticon* (3), 4-20.

Vanderveen, G. (2006). *Interpreting Fear, Crime, Risk and Unsafety*. Dissertatie (mr. N.J.H. Huls & dr. H. Elfers.) Den Haag: Boom Juridische uitgevers.

Citizen journalism, surveillance and control

*Agneta Mallén**

1 Introduction

The mass media today is full of pictures and film clips taken by non-professional–citizen journalists. These video clips and still pictures frequently become 'breaking news' in the international media (Mortensen, 2011). Some argue that the general public first became aware of amateur reporting as a serious matter in 2004, when pictures from the Abu Ghraib prison spurred international debate about the war in Iraq; others claim that the power of amateur coverage was impressed on the general public after 9/11 (Mortensen, 2011). Footage and still photos taken by non-professionals have since repeatedly appeared in the news, especially in such instances as the terror attacks in London in 2005 and Mumbai in 2008, natural disasters such as the tsunami in Japan in 2011, and the revolutions in North Africa and the Middle East in 2011. These forms of participatory publishing are made possible by new communication technology such as the mobile phone, and the emergence of the Web 2.0 technology, especially YouTube (Antony & Thomas, 2010; Rosenberry & St John III, 2010).[1]

The aim of this paper is to examine the consequences of citizen journalism, both for the object under surveillance–in the shape of integrity risks and disintegrative shaming–and for citizen journalists themselves–panopticism, crime prevention, and crime detection–all viewed in terms of control, and by drawing comparisons with the use of surveillance cameras. The analysis is based on the two first known cases of citizen journalism in Finland, from 2006 and 2007, both of which were criminal acts that would have had legal consequences if reported–assault

---

* Agneta Mallén works as Senior Lecturer at the Department of Sociology at Lund University, Sweden, where she teaches criminology. Her primary research interests are crime control, fear of crime, and child abuse and neglect against disabled children. Her recent article on child abuse and neglect has been published in the *Journal of Scandinavian Studies in Criminology and Crime Prevention*.

1. In a survey by the European Commission in 2008, a mean of 24 per cent of the inhabitants of member states used mobile phones (Reuters, 2011). In 2006 (Laurila, 2006) it was predicted that in 2009 some 48 per cent of mobile phones sold worldwide would be camera phones. In 2010, 81 per cent of global sales were of mobiles with a built-in camera (Laurila, ibid.).

and battery, and racist comments that fall in the category of hate crimes. In the first case, a passer-by filmed two security guards assaulting a man. The resulting film clip was posted on YouTube, from where it was picked up by several on-line versions of the Finnish daily newspapers (*Helsingin Sanomat* and *Hufvudstadsbladet*, for example). In the second case, a student filmed a parliamentary candidate making comments with racist undertones. The footage was shown on the television news and was subsequently posted on YouTube. In both cases, the student and the passer-by were practicing citizen journalism, as the film clips were published and discussed online and in the mass media.

2 What is citizen journalism?

The concept of 'citizen journalism' has a variety of meanings, some of them not immediately obvious. So what does the term actually mean?[2] A recent study calls for a broad conception of citizen journalism: one which is not exclusively an online phenomenon, nor confined to 'alternative' news sources (Goode, 2009). In 2004, Dan Gillmor divided citizen journalism into two principal categories.[3] The first comprises citizen journalists with a strong commitment to local politics and decision-making at a local level. They have previously participated by sending letters about local decisions and local politics to various daily newspapers, but now publish in their blogs. To this category I would add 'traditional' citizen journalism, which in English is covered by the term 'public journalism'. The theory of public journalism advocates a form of journalism based on how ordinary people describe reality and problems. Peter Beckman, a journalist for the Swedish daily newspaper *Dagens Nyheter*, writes (2003) that public journalism seeks to treat its readers as possible actors, not merely as passive consumers of news. The public journalism movement emerged in conjunction with the 1988 presidential campaign in the US, with the objective of increasing civic commitment in the democratic process of the presidential elections. The aspiration was to involve larger numbers of people in public discourse and to improve the way in which democracy functioned (ibid.). One of the foremost advocates of citizen journalism is Jay Rosen, professor of journalism at New York University, who describes the emergence of the citizen journalism movement in his book *What Are Journalists For?* (1999).

The second category of citizen journalism comprises the individuals whom Gillmor (2004) termed 'the former audience': listeners, viewers, and readers who

2. I will use the concept of citizen journalism as a generic term for the concepts public journalism, civic journalism, participatory journalism, citizen journalism, and grassroots journalism, although all these concepts represent slightly different kinds of citizen journalism.
3. Gillmor, however, uses the term 'grassroots journalism' in his book (ibid.).

now act as citizen journalists by e.g. blogging or creating websites. According to Gillmor (ibid.), they provide an important source of news for others, including professional journalists. The term citizen journalism is used broadly today, and includes practices such as current affairs-based blogging, photo and video sharing, and posting eyewitness commentary on current events, as well as activities such as re-posting, linking, tagging and modifying, or commenting on news material posted by other users or by professional news outlets (Goode, 2009; Lasica, 2003). Here we use the term citizen journalism in the wider sense discussed by Goode (2009) to analyse the two Finnish cases in which the news material was published both on YouTube and in the 'ordinary' mass media.

The advantages and disadvantages of citizen journalism are of a general nature. The effectiveness of citizen journalism was seen, for example, in the revolutions in North Africa and the Middle East in the spring of 2011, for word of the protests principally spread quickly through social media such as Facebook and YouTube. Another advantage is that the journalistic material created in this way is democratic, as even the humblest members of a community can air their views. Citizen journalism can provide the independent, reliable, accurate, wide-ranging, and relevant information that a democracy requires (Antony & Thomas, 2010). In an article in the Swedish newspaper *Sydsvenska Dagbladet* (Chukri, 2007) on Western European IT companies that, consciously or unconsciously, promote internet censorship in China, citizen journalism and 'citizen paparazzi' are advocated as a means of spreading democracy in China: 'It would also have been a good deed, if the companies that assisted… (among others, Microsoft, Yahoo and Google) donated a few million untraceable camera phones. Citizen paparazzi have to be a good start for a democratic revolution' (ibid.). Gillmor (2004, p. 140) argues along similar lines when commenting on online citizen journalism: 'However, in country after country where free speech is not given, the blogosphere matters in far more serious ways. This is the stuff of actual revolutions.'[4]

The positive aspects of citizen journalism are that this kind of journalism can increase freedom of expression and strengthen democracy. Also, citizen journalists are able to produce information that challenges the official version of events. Citizen journalism thus offers an information-communications marketplace that sustains the commodification and mass consumption of adversarial, anti-establishment news (Greer & Mc Laughlin, 2010).

Criticism of citizen journalism focuses primarily on the fact that its credibility cannot be guaranteed. This form of publishing cannot be edited or subject to edi-

4. The Arab Spring in North Africa and the Middle East in 2011 reveals the prescience of Gillmor's comments. The young people who defied their regimes used blogs, social media, and text messages to spread the revolt.

torial review, and, in the absence of editing, citizen journalist media material can hardly be thought a valid form of journalism, its critics claim (see, for example, Lasica, 2003; Mortensen, 2011. Moreover, because of the informal character of the photos, the veracity of the news cannot be guaranteed. Still, the media concerns that publish citizen journalists' material are said to comply with legislation and regulations, including knowing where and how recordings were made and ensuring that those who recorded the material do not have a hidden agenda.[5]

3 Two cases of citizen journalism

What, then, of the two Finnish cases of citizen journalism that prompted my interest in the subject, with its potential ability to control and monitor the individual? The first case, an assault, resulted in a trial, while the second case, racist utterances, did not have any legal consequences.

Kicking a man when he's down

In October 2006, a man was beaten by two security guards when they arrested him outside a shopping centre in southern Finland. The event attracted considerable attention because a passer-by secretly filmed the assault with a camera phone and posted the footage on YouTube.[6] The clip first shows how one security guard beat the man with a baton before the man was arrested–the man had previously tried to hit the security guard with his fist. What then followed was subsequently heavily debated: the other security guard kicked the arrested man twice in the back, while he lay on his stomach on the ground with his hands tied in handcuffs behind him.[7] This second guard was convicted of assault in June 2007 (Kerkelä, 2007), and was sacked by the security company he was working for when the assault took place. The district court, however, imposed no punishment on the other guard because he and the assaulted man arrived at an out-of-court settlement.

5. These arguments for the correctness of publishing citizen journalist material have been advanced by FST, Finland's Swedish Television, for example, in an online commentary in March 2007 (Medborgarjournalistiken kommer, 2007).
6. The first year the clip was on YouTube it was downloaded over 170,000 times, reaching a figure of 218,000 downloads by March 2011.
7. The first security guard was also interviewed by the police, but was let off. The victim was suspected of assault because he had hit and kicked his girlfriend inside the shopping centre before the guards arrived (Kerkelä, 2007).

Fraudulent journalism

The second event occurred in central Finland during an election rally for the parliamentary elections in the spring of 2007. During a conversation with a voter, a parliamentary candidate made the following comments about Somalis.[8]

Voter:	Can you, like, say that – there's no way I'd want to be a racist but – people who come from African countries, have they adapted better to these problems?
Parliamentary candidate:	There are two good examples… the Asians who came from Vietnam… and then came to this country, so they've adapted very well. … The other extreme case is the Somalis who hang about the central station, using drugs and chewing tobacco, without any intention of learning the language or going to work.[9]
Voter:	What can we do? I still want to know, should we just accept the situation as it is?
Parliamentary candidate:	Vote for [the party], there are no other alternatives, no other options. And we must have the interior minister portfolio!

However, it so happened that a student was listening to the initial conversation between the parliamentary candidate and the voter, and then on an impulse secretly filmed the rest of the conversation with her camera. To show that the conversation was taking place at a public election rally, the student began by filming an establishing shot and then put the parliamentary candidate in the frame. In a later newspaper interview she was reported as being 'shocked' by what she had recorded in the film sequence (Myntti, 2007). Deciding that her material should reach a wider audience, she therefore contacted Finland's Swedish Television (FST) in order to get it released on the television news. A few days later, the filmed sequence was broadcast on *TV-Nytt*, Finland's Swedish-language television news.

In a later comment, made on the day the film clip was broadcast on the national news, the parliamentary candidate stood by his statement, saying that it represented his party's general line (Myntti, 2007). The party chairman denied this and wanted to review the incident. The next day, the parliamentary candidate stated that he was the victim of a political stunt, and used the term 'fraudulent journalism' to describe what he had been subjected to. To circulate the film

8. I have translated this exchange from the Finnish. The conversation between the MP and the voter is available on YouTube at <http://www.youtube.com/watch?v=vzkGDrKD8js>.
9. The parliamentary candidate may be referring to the drug *qat*.

sequence further, the student posted it on YouTube at a later stage.[10] The parliamentary candidate was elected to the Finnish Parliament in the 2007 parliamentary elections. However, the party he represents did not obtain the interior minister portfolio mentioned in the film sequence.

4 Surveillance and control

Consequences for the surveillant

Given the fact that control by citizen journalism is dependent on cameras, one might ask if there is a resemblance to the forms of control exercised by means of surveillance cameras. The latter is often discussed in terms of Foucault's concept of panopticism (1977) (see, for example, De Saulles & Horner, 2011; Koskela, 1999; Koskela, 2003). According to Foucault, panoptic control is invisible and seeks to discipline the individual. His model has lately come in for criticism as being unsuited to the contemporary global, technological, and political dynamics of surveillance (Haggerty, 2006); yet despite this, definitions that are variations on the term panopticon are still frequently used in studies of social control.[11] As Haggerty argues (ibid.), the inability of current research to abandon the term indicates that panopticon now stands for surveillance itself. Panoptic control has since taken on new shapes in neo-panopticons such as CCTV (Mann, Nolan & Wellman, 2003). In public or semi-public locations, individuals are liable to become unwilling and sometimes unknowing subjects of surveillance, and the knowledge that they may be under surveillance may be sufficient to induce obedience to authority. The individuals subject to neo-panopticons do not have direct visual or aural contact with the surveillants. Individuals are controlled by neo-panopticons in two ways: they are subject to observation on video monitors, and they are under the potential control of people in positions of authority. Access to neo-panoptic control is one-sided and undemocratic, as the general public is the object of neo-panoptic control, not in control of it (Mann et al., 2003; Taylor, 2010).

Citizen journalism, with its use of camera phones, represents a third form of control that is dispersed and democratic (Best, 2010; De Saulles & Horner, 2011; Mann et al., 2003). Control has shifted from controlling organisations and agen-

10. During its first year on YouTube the clip was downloaded 45,000 times.
11. In addition to the catopticon (Ganascia, 2010) and neo-panopticon (Mann et al., 2003) used in this article, Haggerty (2006) notes the omnicon, the ban-opticon, the urban panopticon, and the social panopticon.

cies and is now exerted by individuals in their everyday lives (Best, 2010).[12] Ganascia (2010) calls this the *catopticon* (from catoptrics, the study of reflected light) because of its social architecture that permits everyone to communicate with everybody and removes the surveillants from the watchtower. This third form of control has also become participatory, in the sense that surveillants often take part in the actions that they monitor. The control exercised by citizen journalists is sometimes referred to as sousveillance: a bottom-up, inverse panopticon where anyone can take photos or videos of any person or event, and then disseminate the information freely across the world, and where panoptic technology helps individuals to monitor those in authority (Ganascia, 2010; Mann et al., 2003). One can discuss whether in contemporary society these three forms of surveillance–the panopticon, the neo-panopticon, and participatory surveillance–include or exclude one another. It has been argued that at least the two latter forms of control–neo-panoptic control and citizen journalistic control–in fact coexist (Ganascia, 2010).

Citizen journalistic control can be seen as a form of sousveillance (Ganascia, 2010; Mann et al., 2003). This is exemplified by the case of the security guards: the citizen journalist caught two agents of authority–the security guards–on film exercising their power in a way that infringes the law. There are also some strictly panoptic features in the control exercised by means of camera phones: as the individuals under surveillance never know when they are being observed, and must therefore always behave according to the rules, this form of control can to some extent be described as panoptic (see Foucault, 1977). In the example of the parliamentary candidate, there is one panoptic element in citizen journalist control because the parliamentary candidate who is being monitored secretly, and does not know that he is under surveillance. However, citizen journalist control in both cases is not panoptic inasmuch as their control does not seek to impose discipline. Ordinary surveillance cameras are not generally hidden: potential offenders know that they are being monitored and are at risk of being caught; these cameras give rise to a kind of self-control within the individual, which can be likened to Foucault's concept of discipline (1977; Mallén, 2005). That said, there is recent research which argues that citizen journalist equipment–camera phones and the like–can likewise deter individuals from committing crimes (De Saulles & Horner, 2011). In a recent article, De Saulles and Horner (ibid.) argue that individuals may feel pressure to modify their behaviour for fear of surreptitious mobile technology surveillance by strangers. This would certainly correlate with the idea of ordinary, visible security cameras functioning as a deterrent.

12. Koskela (2009, p. 148) argues that the use of surveillance technology has trickled down from the public sector to the private sector and on to private individuals in what he calls the 're-privatisation of surveillance'.

The control that citizen journalists exercise can also be seen as informal social control. Earlier research has presented the idea that surveillance cameras replace informal social control (Koskela, 2000). Koskela (2000, p. 244) argues that 'Electronic means are coming to replace informal social control in an urban environment: *the eyes* of the people *on the street* are replaced by the eyes of surveillance cameras' (my emphasis). Because of citizen journalism, camera phones and similar items transform the eyes of people in the street into miniature surveillance cameras, to borrow Koskela's metaphor. Thanks to the camera phone, the man in the street becomes a citizen journalist who can record unpleasant incidents for dissemination in various news forums. Informal social control has not been replaced by camera control, as Koskela would suggest, but rather camera control has become an instrument of informal social control. Using Mortensen's terminology, citizen journalists have become potential eyewitnesses, mobile cameras at the ready (Mortensen, 2011). Citizen journalists are able to actively record and distribute objectionable or criminal incidents to a wider audience, instead of passively observing such incidents.

Nevertheless, citizen journalistic control is somewhat ambiguous. While surveillance cameras are not aimed at specific individuals but affect all citizens (Listerborn, 2002), citizen journalistic control is aimed at specific individuals. However, it can also affect a larger number of citizens as a deterrent or a means of informal social control. On the one hand, citizen journalistic control can be seen as an expression of net-widening (Cohen, 1985), with more and more people becoming entangled in the control apparatus. Because of the increasing number of camera phones, and thus prospective citizen journalists, the control network is growing. Critics of surveillance technology argue that the control net has become increasingly intrusive: the mesh of the net has become finer, while the net itself has grown (Corbett & Marx, 1991). The mesh becomes even finer when in principle anyone with a camera phone can exercise control over their fellows. On the other hand, citizen journalistic control can be seen as more targeted than the control exercised by surveillance cameras. Not all individuals who have a camera phone or a camera with a built-in video camera in fact become citizen journalists. Those who act as citizen journalists are focused on their activities, and do not film at random everything that comes up in their viewfinders the way traditional surveillance cameras do.

In terms of crime prevention, surveillance camera control has been criticised for being an *ex post facto* solution;[13] surveillance cameras do not prevent all types of crime, and if anything constitute a response to criminality. Citizen journalistic control, like surveillance camera control, can also be regarded as an *ex post facto*

13. Oc & Tiesdell (1997), among others, raised this point in the late 1990s. More recently the critique has been discussed by, for example, Norris, McCahill & Wood (2004).

solution. The control that citizen journalists exercise cannot be seen as crime prevention, while the response to the events is not triggered until the media material is published.

Obstacles to citizen journalistic surveillance

To further broaden the discussion of control, what of the obstacles to identifying and/or reporting individuals filmed by surveillance cameras or citizen journalists' cameras? It is no easy task to identify those who appear in the images captured by 'ordinary' surveillance cameras. The Swedish sociologist David Wästerfors (2006) discusses how complex the process can be when the police set out to identify someone from a surveillance camera image. Successful identification requires that the camera images are sharp and that the police have prior knowledge of the offender. The social aspect is unavoidable in this identification process, as a police officer or security guard will have to look at the images and recognise the person or persons shown, and 'neither "recognition" or "identification" is purely technical' (ibid. p. 39).

Difficulties may also arise when crimes caught on surveillance camera are reported to the police. Surveillance camera control seems to reduce property crimes such as those involving vehicles, theft, and burglary (Koskela, 1999; Wästerfors, 2006), perhaps due in part to a greater willingness to report property crimes to the police, it being a requirement if victims are to receive compensation from their insurance companies. However, studies show that camera surveillance does not reduce violent crime (see Koskela, 1999). The difficulties in reporting crimes caught by surveillance cameras may also depend on reporting procedures: sometimes those involved must themselves report that there are cameras at the scene before the police can download and view the images (Wästerfors, ibid.).

When it comes to identifying individuals filmed, and thus monitored, by citizen journalists, similar problems may arise. Of the two cases described above, it is easy to identify the parliamentary candidate in the film clip on YouTube: the film quality is decent, and his identification is facilitated by the fact that he is a public figure and the election meeting was a public event. In the assault case, the identification of the security guards is harder. This sequence is of much poorer quality than the election meeting footage, and it is difficult to distinguish the guards' faces. Still, as the guards were on duty, wearing their uniforms, it would be a fairly simple task to identify the guards by checking who was on duty the night the incident happened.

It can also be a delicate business, reporting a crime captured on a citizen journalist's film. Returning to the assault by the security guards, it is difficult to say whether it would have come to the notice of the police without the video sequence: the guards would hardly have reported themselves to the police, while their victim may perhaps not have reported the assault either. The guards had arrested the victim because a few minutes earlier he had assaulted his girlfriend, and he can therefore hardly be considered an 'ideal victim' (Christie, 2001) who was behaving himself at the time of the crime.[14] According to the Norwegian criminologist Nils Christie (ibid.), it is ideal victims who most easily gain uncompromised and legitimate status as victims. The assault victim was not entirely guiltless, and he would therefore have found it difficult to assert his innocence and so maintain the guards' guilt. The assault case came to the attention of the police when the security company itself reported the matter to them (Kerkelä, 2007).

Consequences for those under surveillance

Apropos the consequences of citizen journalistic control for those under surveillance, citizen journalistic control is clearly related to issues of integrity and privacy, which have been widely discussed in relation to surveillance cameras (see, for example, Taylor, 2010). The argument against such cameras–that they violate the individual's integrity–has been articulated with respect to the mass media (Koskela, 1999). Surveillance is presented in the mass media as an invasion of privacy, and is coupled with dystopian visions of the future of surveillance societies. This critical view of surveillance cameras, where the cameras that intrude on personal freedom are perceived as more offensive than the crimes they may discourage, was evident in scholarly works as early as 2000 (see, for example, Dubbeld, 2003; Taylor, 2000) and has been advanced many times since (see, for example, Ruegg, November & Clauser, 2004; Sheller, 2008).

The discussion of integrity becomes even more intricate when it relates to the control exercised by means of mobile phone cameras or their equivalent. This is very much the case with the parliamentary candidate: he wrote in his personal blog (Salo, 2007) that although the election meeting in which he made his comments about the Somalis was public, the discussion was wholly private:

> 'Absolutely right…, it was a public election meeting, but our discussion was private. … The fact that our discussion was private seems to be forgotten, and what was done [by the student who filmed it] is defended by saying that the event was public. … It was not a matter of a publicly voiced comment, but an opinion expressed in a discussion between two people. An opinion that has not changed since then.'[15]

---

14. In Finland, security guards have the right to make a citizen's arrest if they catch a criminal red-handed.
15. My translation from the Finnish.

The parliamentary candidate's view of the private nature of the conversation highlights a very interesting facet to camera phone use. He believes that the citizen journalist with her camera crossed the boundary between public and private, and that she violated the parliamentary candidate's integrity by filming his racist comments.

Using Erving Goffman's conceptual framework (2004), one can say that in the case of the parliamentary candidate an outsider unexpectedly entered the 'back region' of a performance in progress. The back region is for Goffman (ibid. p. 120) a private space: when outsiders unexpectedly enter the back region of a performance in progress, it is often the case that the consequences of their untimely entry are best observed not in the effects it has on the ongoing performance, but in those that it may have on a different performance–on what the actors or the audience would normally exhibit to these outsiders at a time and place when the outsiders would have been expected to serve as an audience. In the case of the parliamentary candidate, the citizen journalist is an outsider who witnesses a performance not intended for her. Citizen journalism, as in this case, can also transform a back region into a front, or public, region. While mingling with voters, the parliamentary candidate let his mask slip without realising that, by virtue of being filmed by the citizen journalist's 'hidden camera', he was still in the front region. The parliamentary candidate may even have perceived himself and the voter as two people who shared similar views on immigrants, and who were thus collaborating in the presentation of a performance. People who collaborate in a performance display a familiarity when no audience is present (see Goffman, 2004). When the voter said in the initial question 'there's no way I'd want to be a racist but', the parliamentary candidate may have interpreted this as an endorsement of his own opinions.

The example of the parliamentary candidate also raises the question of the individual's ability to check material before it is made public on YouTube and the like. As Taylor (2010) argues, the explosion of social networking sites such as YouTube makes the individual's control over what is made public even more important. Defining the boundary between private and public material on social networking sites has thus become of the utmost importance to the individual. In Taylor's study (ibid.), schoolchildren viewed very negatively the fact that private photographs were made public on social networking sites without their explicit permission, although they had previously considered the pictures as acceptable for public consumption. Individuals photographed by citizen journalists seldom have the opportunity to check the material before it is broadcast, and therefore feel they have lost control over the media material in which they are involved. The parliamentary candidate's words about the privacy of his discussion with the voter indicate that he and the citizen journalist's views differed on where the boundary between public and private lay. The parliamentary candidate therefore felt he had lost control of the situation.

The individual's lack of control over media material that is made public may also create a stigma (Goffman, 1986). Traditionally, a stigma arises in personal, body-to-body, interaction (Pelkonen, 2011); here individuals have some control over the information spread about their stigma, for example, for holding divergent religious views, for in a body-to-body interactive situation, they are aware that they are being observed and are therefore able to control the information output regarding the stigma. In mediated interaction not of a body-to-body kind, such as that between citizen journalists and the objects of their surveillance, individuals are unable to control stigmatising information.

This stigmatisation is especially vivid in the case of the parliamentary candidate. The racist utterance itself was a deviant act that, undisclosed, would have had few, if any, repercussions. However, the fact that the student filmed the incident and thus disclosed it, making it available to a much wider audience, facilitates the labelling of the candidate's deviant act (Becker, 1963). The citizen journalist made it possible for people outside the dialogue–a dialogue the candidate later defined as private–to view the incident. The party chairman, presumably an important figure for the candidate, also labelled the incident as deviant when he publicly declared that the remarks did not reflect the party line. The party chairman's point of view can also be viewed as a sign of shaming (Braithwaite, 1989). The shaming is disintegrative, not reintegrative, for the deviant act, despite being the target of shaming, does not prick the conscience in a way that might deter later deviancy (as reintegrative shaming would) (ibid.). The parliamentary candidate did not change his deviant behaviour–his objectionable opinions–and in his blog instead asserted that he persisted in his point of view.[16]

5 Discussion

The aim of this paper was to analyse the consequences of citizen journalism for both surveillants and the surveilled. The cases of citizen journalism discussed here are taken from a Finnish context, but can be seen representative of a new

16. Dennis (2008) discusses a similar shaming feature of citizen journalistic control in relation to a case in South Korea, when a young woman refused to clean up the mess when her dog defecated on the floor of a subway carriage. A co-passenger took a photograph of the girl, the dog, and the mess and posted it on a popular website. The girl was soon identified and her personal details were published. According to Denis (ibid. p. 351), this was an attempt to punish the girl for her 'offence' in the form of social humiliation.

form of control that is universal. Citizen journalists' tools–mobile phones and Web 2.0 technology– are used worldwide.

This study shows that the form of control exerted by citizen journalism is exerted by individuals in their everyday lives. The consequences for surveillants centre on panopticism and crime prevention: there are panoptic features of citizen journalistic control because surveillants can exert this control at any point without the surveilled being aware of it. Compared to ordinary surveillance cameras, citizen journalists' cameras exercise a control that is far more targeted. Individuals who have a camera phone do not film randomly in the way surveillance cameras do. However, citizen journalists' camera and ordinary surveillance cameras do share one feature in this respect: neither prevents crime because both constitute a response to criminal actions that is not activated until the media material is broadcast. The consequences for the surveilled involve issues of privacy, integrity, and disintegrative shaming. Citizen journalism can turn a back region into a front region (Goffman, 2004) and dissolve the boundaries between private and public. In addition, citizen journalistic control can be seen as labelling: an act is labelled deviant when disclosed by the citizen journalist and displayed on social media.

There are other sides to citizen journalistic control that are worth considering. As earlier studies have shown, the chance to act as citizen journalists has changed the role and self-perception of the audience (Mortensen, 2011). The individual who was once part of the audience, and who only received information, is now able to act as a citizen journalist, not only receiving but also transmitting information. However, the audience has come to have an increasingly important role in citizen journalistic control in another respect, and further analysis will be required in order to gauge whether citizen journalists both exercise and mediate control. As we have seen, citizen journalists record incidents that audiences then labels as deviant, which would indicate that citizen journalists enable the audience to control the monitored individual, instead of the control being solely exercised by citizen journalists themselves.

This leads to the interesting issue of whether citizen journalists can indeed be seen as capable guardians. As one of the standard situational crime prevention objectives is to increase the presence of capable guardians (Sahlin, 2000), one can question whether citizen journalists can be viewed in this light.[17] In the material

17. The concept of capable guardians is found in Cohen and Felson's routine activity theory (1979), according to which three requirements must be met if a crime is to be said to have occurred: there must be a perpetrator with a motive; a suitable victim; and the absence of capable guardians. The term 'capable guardians' refers to any person who the potential perpetrator perceives as capable and willing to prevent the crime. According to Cohen and Felson (ibid.), the probability of a crime taking place is a function of the probability that these three factors will coincide in time and space.

analysed in this paper, citizen journalists do not directly intervene in what they film: the criminal acts discussed above are neither interrupted nor prevented by the citizen journalists concerned. However, a potential perpetrator may still regard a citizen journalist as a capable guardian simply by virtue of their being present at a criminal event, witnessing and recording it. Furthermore, citizen journalists draw their audience's attention to wrongs and abuses, and may possibly forestall the recurrence of such incidents. If so, this would make citizen journalists capable guardians in a wider sense.

In the words of De Saulles & Horner (2011, p. 214), in the era of citizen journalistic control 'there is nowhere to hide'. However, citizen journalists' control still seems a milder alternative to the rigorous, officially sanctioned scrutiny to which we are subject in a surveillance society. Although all of us who have a camera phone can monitor our fellow human beings, we do not.

References

Antony, M. G., & Thomas R. J. (2010). ''This is citizen journalism at its finest''. YouTube and the public sphere in the Oscar Grant shooting incident. *New Media & Society*, *12*(8), 1280–1296.

Beckman, P. (2003). *Riv stängslen. Medierna som mötesplats: Public journalism i svensk tappning*. Stockholm: Norstedts.

Becker, H. (1963). *Outsiders: Studies in the sociology of deviance*. New York: The Free Press.

Best, K. (2010). Living in the control society: Surveillance, users and digital screen technologies. *International Journal of Cultural Studies*, *13*(3), 5–24.

Braithwaite, J. (1989). *Crime, shame and reintegration*. Cambridge: University Press.

Christie, N. (2001). Det idealiska offret. In Malin Åkerström & Ingrid Sahlin (Eds.), *Det motspänstiga offret* (pp. 46–62). Lund: Studentlitteratur.

Cohen, S. (1985). *Visions of Social Control*. Cambridge: Polity Press.

Corbett, R., & Marx, G. T. (1991). Critique: No soul in the new machine. Technofallacies in the electronic monitoring movement. *Justice Quarterly*, 8, 399–414.

De Saulles, M., & Horner, D. S. (2011). The portable panopticon: Morality and mobile technologies. *Journal of Information, Communication & Ethics in Society*, *9*(3), 206–216.

Dubbeld, L. (2003). Observing bodies. Camera surveillance and the significance of the body. *Ethics and Information Technology*, 5, 151–162.

Foucault, M. (1977). *Discipline and Punish: The birth of the prison*. London: Penguin.

Ganascia, J.-G. (2010). The generalised sousveillance society. *Social Science Information*, *49*(3), 489–507.

Greer, C., & Mc Laughlin, E. (2010). We predict a riot? Public order policing, new media environments and the rise of the citizen journalist. *British Journal of Criminology*, 50, 1041–1050.

Gillmor, D. (2004). The former audience joins the party. In D. Gillmor (Ed.), *We The Media* (pp. 136–157). Sebastopol, Calif.: O'Reilly Media.

Goffman, E. (2004). *Jaget och maskerna. En studie i vardagslivets dramatik* (S. Bergström, trans.). Stockholm: Norstedts. (Originally published 1959.)

Goffman, E. (1963/1986). *Stigma: Notes on the management of spoiled identity*. Harmondsworth: Penguin.

Goode, L. (2009). Social news, citizen journalism and democracy. *New Media Society, 11*(8), 1287–1305.

Haggerty, K. D. (2006). Tear down the walls: On demolishing the panopticon. In D. Lyon (Ed.), *Theorising surveillance. The panopticon and beyond* (pp. 23–45). Devon: Willan Publishing.

Koskela, H. (1999). *Fear, control and space. Geographies of gender, fear of violence and video surveillance.* Helsinki. Helsingin Yliopiston maantieteen laitoksen julkaisuja A 137. Helsinki: Tummavuoren kirjapaino.

Koskela, H. (2000). 'The gaze without eyes': Video surveillance and the changing nature of urban space. *Progress in Human Geography, 24*(2), 243–265.

Koskela, H. (2003). 'Cam Era' – the contemporary urban panopticon. *Surveillance & Society, 1*(3), 292–313.

Koskela, H. (2009). Hijacking surveillance? The new moral landscapes of amateur photographing. In K. F. Aas, H. O. Gundhus & H. M. Lomell (Eds.), *Technologies of insecurity. The surveillance of everyday life* (pp. 147–167). Abingdon: Routledge-Cavendish.

Listerborn, C. (2002). *Trygg stad. Diskurser om kvinnors rädsla i forskning, policyutveckling och lokal praktik.* Gothenburg: Chalmers tekniska högskola.

Mann, S., Nolan, J., & Wellman, B. (2003). Sousveillance: Inventing and using Wearable computing devices for data collection in surveillance environments. *Surveillance & Society, 1*(3), 331–355.

Mallén, A. (2005). *Trygghet i skärgårdsmiljö. En studie om rädsla för brott i Åboland.* Diss., Lund University, Department of Sociology.

Mortensen, M. (2011). When citizen photojournalism sets the news agenda: Neda Agha Soltan as a Web 2.0 icon of post-election unrest in Iran. *Global Media and Communication, 7*(1), 4–16.

Norris, C.; Mc Cahill M., & Wood, D. (2004). Editorial. The growth of CCTV: a global perspective on the international diffusion of video surveillance in publicly accessible space. *Surveillance and Society*, 2, 110–135.

Oc, T., & Tiesdell, S. (1997) *Safer city centres. Reviving the public realm.* London: Paul Chapman.

Rosen, J. (1999). *What are journalists for?* New York: Vail-Ballou Press.

Rosenberry, J., & St John III, B. (2010) (eds.). *Public Journalism 2.0: The promise and reality of a citizen-engaged press.* New York: Routledge.

Ruegg, J., November, V., & Clauser, F. (2004). CCTV, risk management and regulation mechanisms in publicly used places: A discussion based on Swiss examples. *Surveillance and Society*, 2, 415–429.

Sahlin, I. (2000). *Brottsprevention som begrepp och samhällsfenomen.* Lund: Arkiv förlag.

Sheller, M. (2008) Mobility, freedom and public space. In Sigurd Bergmann & Tore Sager (eds.), *The ethics of mobilities. Rethinking place, exclusion, freedom and environment* (pp. 25–38). Aldershot: Ashgate.

Taylor, E. (2010). I spy with my little eye: the use of CCTV in schools and the impact on privacy. *The Sociological Review*, 58(3), 381–405.

Wästerfors, D. (2006). *Kamerans konsekvenser. En fallstudie av kameraövervakning mot bilbrott i Landskrona.* Network for Research in Criminology and Deviant Behaviour at Lund University. Research Reports 2006:1.

Internet sources

Chukri, R. (2007, August 31). Kina–landet med växtvärk. *Sydsvenska Dagbladet*. Retrieved August 31, 2007 from <www.sydsvenskan.se/kultur-och-nojen/kultur_bocker/bokrecensioner/article262930/Kina—landet-med-vaxtvark.html>.

Kerkelä, L. (2007, June 21). Kontulan ostokeskuksen entinen vartija sai tuomion pahoinpitelystä. Käräjäoikeus ei silti tuominnut rangaistusta. *Helsingin Sanomat*. Retrieved June 22, 2007 from <www.hs.fi/tulosta/HS20070621SI1KA02t9e>.

Lasica, J.D. (2003, August 7). What is participatory journalism? *Online Journalism Review* Retrieved April 23, 2007 from <www.ojr.org/ojr/workplace/1060217106.php>.

Laurila, J. (2006, November 2). Gartner: Puolessa maailman kännyköistä kamera. *Tietoviikko*. Retrieved September 20, 2007 from <www.tietoviikko.fi/doc.te?f_id=1057739>.

Medborgarjournalistiken kommer. (2007). Retrieved March 20, 2007 from <http://internytt.yle.fi/bakgrundsartikel.php?id=94467&tema=134>.

Myntti, K. (2007, March 14) Hon smygfilmade Petri Salo. *Vasabladet*. Retrieved September 20, 2007 from <www.vasabladet.fi/popup/print.aspx?storyID=2933>.

Pelkonen, M. (2011) *Informaatio- ja kommunikaatioteknologian rooli sosiaalisen stigman käsitteessä*. Master's Thesis, Jyväskylä University. Retrieved October 2, 2011 from <http://urn.fi/URN:NBN:fi:jyu-2011062311053>.

Quarter of EU homes use mobile phones only: survey (2011). Retrieved October 12, 2011 from <http://uk.reuters.com/article/2008/06/29/tech-eu-telecoms-dc-idUKL2727442920080629>.

Salo, P. (2007, March 14) *Vaasan veri ei vapise*. Retrieved September 20, 2007 from <www.petrisalo.net/blogi.section/pages/blogi_tark81a0da499.html>.

The nodal-network fallacy in the surveillance of transit migration in Belgian harbours

*Marleen Easton and Fien Gilleir**

1 Introduction

As a general introduction to this chapter it may be interesting to explore the website of www.transitmigration.org. The MigMag project develops a new form of cartography on the basis of a dynamic data stock and with the help of IT tools. The cartography represents the migration and border regime of the EU, which is not seen as fixed but as a social and discursive space of negotiation. Interesting is that the visitor of this website can explore how the various key players in the public and private spheres are interrelated and funded, as well as the ways in which these spheres overlap or differ in terms of focus, location or personnel. One can explore five maps that illustrate the actors, discourses, policy documents, and places and practices. The ports can be found in the map 'places and practices' as one place which is of interest to the phenomenon of transit migration.

This contribution zooms in on one aspect of the broader picture sketched above. The seaports of Ostend and Zeebrugge may be seen as geographical 'nodes' in a multitude of flows (of people, goods, information, etc.) as defined by Castells in his work 'the rise of the network society' (Castells, 1996). Within this 'network society', global evolutions take place that have implications for some locations and their security mechanisms. The phenomenon of transit migration and its implications for the question of security in these Belgian seaports is presented here. This contribution does not focus on those who are the subject of surveillance but focuses on those who carry out (social) surveillance of transit migrants who enter Belgium via the seaports of Ostend and Zeebrugge. In this way, the impact of a

* Marleen Easton is professor at the University College Ghent – Faculty of Business and Public Administration where she is director of the 'Governing and Policing Security' research unit (www.gaps-ugent.be), Guest professor at the Faculty of Economics and Business Administration of Ghent University and visiting professor at Rühr Universität Bochum, Master of Criminal Justice, Governance and Police Science (www.macrimgov.eu).
Fien Gilleir is a PhD student at Ghent University, Research Unit 'Governance of Security' (GofS) and researcher at the Research Unit 'Governing and Policing Security' (GaPS) at Ghent University College.

global phenomenon (international migration) on the governance of security within a local entity (the port as a node) is reconstructed from the perspective of those involved. The logic they employ, the tasks to which this gives rise, and their collaboration is explained on the basis of empirical research. Concepts such as nodal governance, nodes, networks and nodal orientation are used as a framework for this qualitative study consisting of interviews and the study of documents.

This article is divided into three parts. In the first part, we explore the relationship between concepts such as 'flows, nodes and networks', 'nodal orientation' and 'nodal governance', which tend to suffer from a lack of clarity in the field. In the second part, we outline the phenomenon of transit migration in Belgian seaports and look at the contextualisation of the phenomenon itself (Jordan & Düvell, 2002, Van Poucke, 2004). We describe the seaports as geographical nodes in a network society and map the various central actors in relation to the phenomenon of transit migration. Finally we reflect on their main logic and on the extent the actors being identified form a network. In the third part, we conclude with one or two reflections on the so-called 'nodal-network fallacy' we have encountered.

2 Conceptualisation of flows, nodes, networks, nodal orientation and nodal governance

It is appropriate to examine closely how concepts such as flows, nodes, networks, 'nodal governance' and 'nodal orientation' are actually developed in a number of different contexts. Various interpretations and applications often display less common ground than might be suspected at first sight. Therefore concepts are not mutually interchangeable. What, for example, is known in the Netherlands as 'nodal orientation' (or 'nodal policing', Bekkers & Van Sluis, 2009), is not the same as the term 'nodal governance' as interpreted and developed by Wood (2006). In what follows, we examine each of these central concepts, paying attention to their definition, implications for the issue of security and their relationship to one another. It is, however, not our intention to be exhaustive in terms of interpretations and applications being made around the world, and examples are merely used in order to clarify.

Flows, nodes and networks

The term 'node' is derived from the Latin 'nodus' (Latin for 'node', 'intersection'). The term 'nodal' is attributed to the Catalan sociologist Manuel Castells, who published, at around the turn of the century, a voluminous and groundbreaking triptych, 'The Information Age' (Castells, 1996). In the first book of this sizeable oeuvre – 'The rise of the network society' – he devotes a section to the 'nodal' outlook, a view according to which large-scale transport movements can be equated with flows that converge – intersect – in nodes (critical infrastructures). This reflects Castells' image of a society, the structure of which has altered over time to become a 'network structure'. This conviction is based on the finding that all kinds of social changes have occurred in the past which have meant that, over time, society has developed into a complex network of movements within which all sorts of branches and nodes form. He bases the rise of information flows on the almost endless possibilities in the field of information technology and the almost unstoppable advance of new technologies that have appeared. A technological revolution has occurred in practically every area of life, such as work, leisure and healthcare. In short, the arrival of this era of information technology has meant that more and more virtual flows are beginning to take shape.

The rise of these virtual flows has at least two major consequences. First, the advance in technological innovation has meant that information has constantly gained in importance. Thanks to technological ingenuity, the exchange of information or data no longer requires people to physically meet to exchange that information/data. Henceforth, this exchange can also take place virtually. The omnipresence of these virtual flows generates a second consequence. Castells claims that the emergence of a 'virtual' reality has repercussions for the social structure of society because the significance henceforth given to places is now being interpreted differently. By using virtual networks, the significance of places has changed considerably over the years, creating a dialectic contrast between 'space of places' and 'space of flows'. This implies, among other things, that society has assumed the form of a network structure, in which certain nodes have constantly become more explicitly visible, within a complex network of national and international connecting circuits.

It was against the background of all kinds of social changes that Castells came up with a number of innovative ideas about flows and nodes in the network society, concepts which he used as an approach to study the new global geography and globalisation. Based on his assumptions, we can say that, globally, there are series of interactions that arise between geographically remote social actors. This creates a network of flows. Essentially, this means that what is important is no longer distances, but whether or not one is part of a network in which functions and processes are organised. Such a network, according to Castells, contains 'nodes', places where key functions within a network are performed and connect with one another. These may be both physical and virtual, for example, an airport or social media such as Facebook. In short, as a result of globalisation, through the rise of information technology and the like, society has been gradually transforming into a network society, an image that Castells uses to point to social developments, including developments in crime, such as organised crime and cyber crime.

The question that arises is what implications a network society (in terms of flows, nodes and networks) has for supervision and enforcement in relation to current security issues. In the light of a number of tragic events (the terrorist attacks of 9/11 being the most salient example), there is a growing realisation that the nature of the new threats is assuming an increasingly transnational character. The significant increase in the magnitude of movements raises questions in relation to the exercise of control, enforcement and supervision. In the aftermath of terrorist attacks people across the globe have become aware of the vulnerability of certain nodes such as airports, seaports, stations and virtual banking. Flows converge on these 'critical infrastructures' and they form major veins of a nation state because of their massive economic potential. The transnational character of the nature of new threats therefore prompts a rethinking of security since supervisory authorities, generally local, are on shaky ground with threats of a more global nature. Therefore, what Castells is driving at, is that if we assume that global flows have become the dominant spatial manifestation of their kind in today's world, then the question arises as to the extent to which the existing 'geographically limited' organisation of security is still able to cope with transnational phenomena.

Nodal orientation

The Netherlands in particular has been keen to apply Castells' thinking. In 2005 a policy document 'Police in Development' (*'Politie in Ontwikkeling'*) was generated under the auspices of the Board of Chief Commissioners (*Raad van Hoofdcommissarissen*) (Bakker, 2006; Boutellier, 2007). This document included a shift in policy from a local to a 'nodal orientation' for the Dutch police and the concept of 'nodal police' was introduced (Bekkers & Van Sluis, 2009; Tol, 2007).

Both the concepts 'nodal orientation' and 'nodal police' contain the reasoning that an increase in the scale of social activities has led to the increased mobility and flexibility of information, money, goods and people. Since all of these flows move from place to place, they also pass along infrastructures such as airports, seaports, stations and shopping or business centres. The policy document focuses heavily on these kinds of specific geographical areas and physical infrastructures which contain different flows and hold significant social, economic and political capital. The need for advanced and increased supervision, enforcement and detection on these infrastructures is stressed (Bekkers & Van Sluis, 2009).

This shows that the Dutch approach is therefore largely based around the functioning of the public police. The matter is heading explicitly in the direction of

an extension of the powers of the authorities (and the police in particular) to make inquiries and carry out checks. At the same time, the global focus falls mainly on the proactive deployment of police capacities on the one hand and the importance of information on the other. The 'nodal orientation' therefore mainly resembles a network approach to deviant behaviour on and along infrastructural nodes. The abstract sociological concept of Castells is thus operationalised in favour of police practices (Ferweda, Van Bolhuis & van der Torre, 2008, p. 21).
While acknowledging that flows move globally, these policy-makers also realised that the emergence of new technologies, together with the disappearance of physical borders, have led to anonymity in the flows and therefore to invisibility of behaviour. The decline in opportunities for governments to act on these anonymous flows then translates itself into the way in which the Netherlands wants to deploy its enforcement mechanisms in the fight against new (and large-scale) forms of crime. This places the focus squarely on the unmasking the bad as a result of all manner of trends that have led to expansion, increased mobility and anonymity. The best way to achieve this is, on the one hand, by using hi-tech means and, on the other, by granting the police extensive powers. A striking example of this is Automatic Number Plate Recognition (ANPR, also known as CatchKen) introduced several years ago in some cities in the Netherlands. The aim of this system is to monitor all flows of passing vehicles. It compares all number plates with various databases. Through proactive camera recording, the authorities want to combat dishonest behaviour in flows and, where possible, identify it at an early stage. The main goal of these technologies is to generate visibility of all users in any of the flows mentioned above. It is expected that within a few years it will be hard to travel without being seen by these new technologies.[1]

Inherent to the concept of 'nodal policing' in the Netherlands is the view that it is not (any longer) possible for the police to protect society on their own. In particular, this is based on the assumption that the police must cooperate, i.e. network, with partners both inside and outside the 'security chain'. In this way 'nodal policing' transcends the notion of 'nodal orientation' and relates to the concept of 'nodal governance', which is elaborated on below (Burris, Drahos & Shearing, 2005).

1. Charlotte Van Ooijen, working on her PhD on 'The implications of location aware technologies (LATs) for the government-citizen relationship' at the University of Tilburg, in an interview (28 March 2011) on http://www.wittekentekenplaat.nl

Nodal governance

The concept of 'nodal governance' covers a different type of approach to the concept of 'nodal orientation'. In contrast to 'government', which refers to a system of centralised political authority with the public police as primary agent of security, 'governance' alludes to a fragmented way of conducting policy (Shearing & Wood, 2003). In the 'governance of security', both government and extra-state actors are involved at sub-national, national and international level in addressing security questions. Typical of this switch from 'government' to 'governance' is the emergence of a network character in which the idea is abandoned that it is by definition the responsibility of the government to assume a leading role in fighting crime phenomena. 'Today, the police are only one node in a network of auspices and providers of nodes that work to govern security both alone and in conjunction with each other.' (Shearing, in: Bakker, 2007). From this reasoning, actors are seen as nodes, and it is not so much about physical infrastructures like those that are central to the concept of 'nodal orientation' outlined above.

From the perspective of 'nodal governance' networks are understood to be sets of actors/nodes that share a common interest in a particular field and can be linked together by formal or informal relations (Wood, 2006). It is precisely by analogy with so-called criminal or 'dark networks' that exert power by means of decentralised nodes and through information flows that Shearing & Wood (2003) argue in favour of developing 'bright networks' that can also be put together flexibly and thus adapt quickly to changing circumstances. This implies that less obvious nodes – actors outside the security domain whose existence is often denied – also have to be mobilised in order to generate new intelligence.

Our central argument is that the concept of 'nodal governance' can be considered to add a crucial dimension to the notion of 'nodal orientation' in our study of the surveillance of transit migration in Belgian harbours. Apart from the valuable notion of geographical areas and physical infrastructures as 'nodes' in a networked society, 'nodal governance' adds to the notion of state and non-state actors being 'nodes' and raises the question of whether there is a kind of network between the actors at 'play' in the field. Furthermore, the 'nodal-network fallacy' that Shearing & Johnston (2010) refer to captures the idea that the presence of 'nodes' does not necessarily imply the existence of a 'network'. This is a pitfall when trying to interpret what is actually going on in the field. Window-dressing should be uncovered and explanations should be unravelled on the basis of empirical research in the field.

As Shearing & Johnston (2010) aim for empirical analyses in which 'nodal mapping' on the one hand and 'network analysis' on the other are carried out, we used this framework in our empirical research in the Belgian harbours. The 'nodal mapping' consists of a first inventory of the relevant actors, 'nodes', which can be linked with the phenomenon. Using cartographic mapping one is able to draw a picture of who is involved: what actors can be situated within the perimeter and what is their original goal? It refers to how 'nodes' think about security (the mentalities); the methods they might use to facilitate it (technologies); the social, cultural, economic or other means they might deploy (resources); and the structures that enable the mobilisation of resources, mentalities and technologies in pursuit of the security problem being addressed (institutions). From this approach, a 'node' brings together knowledge, skills and resources. A sub-question is whether any nodes/actors are ignored in addressing these issues, taking into account that state and non-state actors can play a role.

The 'network analysis' is then directed at answering the question of whether or not the various actors in the field are cooperating. To phrase it differently, the question is whether there are mutual relations and, if so, whether these are characterised by cooperation or competition. This makes the question of whether nodes are networked a matter of empirical enquiry (Johnston & Shearing, 2003).

In summary, the 'nodal governance' perspective provides a frame for analysing developments in the light of new security settings. It generates a focus on the 'pluralisation' of actors, with an eye for the wide range of governmental as well as non-governmental 'nodes' that organise themselves around a phenomenon, assessing the extent to which they form a network. This is the conceptual framework within which we shall study below the 'governance of security' in relation to transit migration in Belgian harbours.

3 The surveillance of transit migration in Ostend and Zeebrugge

Contextualising the phenomenon of transit migration in Ostend and Zeebrugge

In an attempt to escape war, oppression, natural disasters or poverty (so-called push factors), people are moving around the globe. The increase of immigration to Europe is just one part of this global mobilisation (Van Houtum & Pijpers, 2005). Since the 1985 Schengen agreement allows EU-citizens to move and reside freely within the EU itself,[2] European borders (and Schengen borders in particular) are closely monitored for movements from non-EU countries. In Belgium borders are monitored at the international airport of Brussels, the regional air-

2. Regulation (EC) No 562/2006 of the European Parliament and of the Council of 15 March 2006 establishing a Community Code on the rules governing the movement of persons across borders (Schengen Borders Code).

ports of Deurne, Ostend, Wevelgem, Gosselies and Bierset, the seaports of Antwerp, Ghent, Ostend and Zeebrugge and the Eurostar-terminal in Brussel-South railway station. These ports are undoubtedly logistical nodes where inland waterways and railways converge. Most are also close to main roads or motorways, and can thus be reached quickly and easily. One side-effect of this ready accessibility is that even unwanted visitors can effortlessly use this easy access. After all, car parks are also used as locations where people can climb into lorries which then head to the seaports. In addition to these important functions in connection with maritime traffic and logistics, a lot of these cities (Brussels, Antwerp, Ghent, Ostend and Zeebrugge) also have major tourist attractions. This generates an intersection of legal and illegal movements.

Given their strategic location, it comes as no surprise that the Belgian seaports of Ostend and Zeebrugge, central in this contribution, form a major hub for people with an illegal status 'passing through'[3] en route to the UK. Belgium is a major traffic node thanks to the various traffic arteries from the surrounding countries that converge on the E40 heading for the coast. The Belgian coast acts as a bottleneck for the movement of migrants towards the UK.[4] Add to this the many ferry links that make several crossings a day to the UK, and the use of this mode of transport by transit migrants is obvious.[5] According to the statistics produced by the Waterway Police in Ostend and Zeebrugge, they have been intercepting around 200 people each month, a number that tends to keep on rising. Most of these people come from India, Afghanistan or Algeria (Heymans, 2008, 2009). The UK seems to be their dream final destination where they believe they are welcome to build a new life. In September and October 2011, the local police of Ostend arrested almost 300 transit migrants.[6] Research shows that controls have not led to a decrease in illegal migration but instead have strengthened the efforts of migrants to avoid confrontations with law enforcement agencies in order to reach their final land of destination (Cyrus & Vogel, 2002; Jordan & Düvell in Düvell et al., 2002, p. 85). The control of transit migrants in the harbours of Ostend and Zeebrugge are a priority as transit migrants are seen to be a threat to the quality of life of the residents and tourists of Ostend and Zeebrugge. It is clear that this phenomenon of transit migration is an issue for Ostend and Zeebrugge with considerable social relevance (Van Meeteren, Van San, & Engbersen 2007, De Bie, 2005; Heymans, 2008, 2009).

3. We consistently use 'people with an illegal status' instead of 'illegals' as we agree on the argument that a person as such cannot be illegal but their administrative status can.
4. See in this connection Federal Police, 'Transitland België', *Infodoc* December 2007, 1-5.
5. *P&O Ferries hervat vrachtroute Zeebrugge-Dover vanaf 7 april* (P&O Ferries to reopen shipping route Zeebrugge-Dover starting 7 April), De Morgen newspaper, 23/03/2010.
6. www.deredactie.be, 26/10/2011.

Nodal mapping & network analysis of the surveillance of transit migrants

The question arises as to what is actually going on in relation to the surveillance of transit migration in the seaports of Ostend and Zeebrugge. Nodal mapping comprises a set of questions such as 'who?', 'what?', 'where?' and 'when?'. The main questions addressed here are 'Who are currently the actors that are active in relation to this phenomenon?' and 'What is their logic?' A network analysis is done through questions related to the perception (reciprocal image) of who is performing the directing role and the extent to which interrelations are of a formal nature. We also gauged opinions concerning the commitment of other actors: respondents were asked whether there was any mutual cooperation and, if so, what its dynamic was.

In what follows, our main findings are presented. We rely on a five-month empirical study commissioned by the Waterway Police that was completed in the spring of 2010 (Easton & Gilleir, 2010).[7] We used a triangulation of methods, consisting of the study of documents and in-depth interviews with 15 actors who are faced with this issue, either directly or indirectly, in the course of their work. This comprises a rather diverse group of actors which includes the Waterway Police, local police, federal police, harbour master services, Immigration Department, Department for Domestic and Border Controls (FPS Interior), Guardianship Service (FPS Justice), the Coast Guard and various centres for social welfare in the vicinity of the harbours.[8] The selections of actors within the social welfare sector was the result of an active search for actors that play or could play a role in tackling this phenomenon in ways that differ from those of the authorities such as the police. In what follows, we illustrate the most obvious logic in relation to the most important actors in the field. This reflects our empirical research in a nutshell and illuminates the confrontation of logics in the field.

Security logic within the Waterway Police

> *'Illegal migration is a security problem: people die during the crossing. We had 58 dead Chinese here in 2000, then we had the Wexford drama. This is a security problem, and we are the authority responsible for security in the port.' (Interview officer Waterway Police).*

7. For earlier work on the nodal governance in Belgian harbours see Gilleir, F., Easton M., Ponsaers, P. & Cools, M. (2008), 'Nodale politie in de Antwerpse haven', unpublished research report commissioned by the Police Science and Research Programme, the Netherlands.
8. Social organisations such as Intake Centre Kwiedam, Centre for Advice for Immigrants (non-profit organisation), Outreach Work (*Straathoekwerk*) Ostend, Jakoeboe (non-profit organisation); Social Welfare Centre Refugees etc.

As a rule, it is the Waterway Police that comes into direct contact with transit migrants through its frontline role in the port area. This brings with it a heavy workload (Vanoverschelde, 2007). After a security search, an administrative check is carried out with the aim of establishing the identity of the person concerned. Fingerprints are taken, together with an iris scan that is used to support identification. In many cases, a bone scan will also be used in an attempt to reveal whether the migrant is a minor or an adult. The person is then given a questionnaire in the hope of gaining an insight into the motives behind the journey, and possible links with smugglers are also investigated. On an administrative level, therefore, a great deal of work has to be carried out under the auspices of the front-line police services. Based on an administrative report, the FPS Immigration is subsequently contacted in each case; they then make a decision on the further fate of the person concerned.

At the same time, medical help is required in most cases, and people generally need to be fed. Given the often neglected physical condition of the people concerned, the Waterway Police finds itself obliged to handle the transport to hospitals, reception centres, etc. at its own expense. In other words, the security logic is besides the prevention of crime also a function of guaranteeing the physical integrity of the migrants passing through. However, the resources and infrastructure available to the frontline police service to temporarily accommodate people are very limited. There is, however, a separate area where illegals can be kept for a short time. In this connection, a creative approach was adopted in the buildings of Zeebrugge Waterway Police, and the police inspectors' personal changing rooms were converted into an 'Administrative and Judicial Handling of Illegals Area'. Consequently, the Waterway Police developed a clear humanitarian touch to their approach. Nevertheless, the treatment of transit migrants is made difficult by language problems and cultural differences.

The Waterway Police invests a great deal of capacity in dealing with these cases, but the large numbers of transits, arrangements concerning meals and care mean that the deployment of staff is at the expense of core tasks for which the Waterway Police is legally responsible. Respondents report a daily confrontation with 15 to 40 people with an illegal status during their working day. The local police are in fact endlessly getting people off the street who are released again and get themselves ready for another attempt to flee to the UK. The police officers regard an approach of this kind as fighting a losing battle as no overall policy is dealing with this hopeless situation. As a consequence, the lack of an immediate solution and the high working pressure, in combination with limited capacity, lead to a decrease in motivation within the organisation of the Waterway Police. Until now, several attempts to share the efforts and to engage other partners has led to nothing. Services such as the Immigration Department, the federal police and other governmental agencies are reluctant to take on more responsibility than is legally prescribed. They all strongly limit their activities to their official 'core tasks', which are border control and investigative work.

Economic logics within the port

> *'It is extremely important that you have a reasonably safe port, as this is more attractive to customers. Everything that goes on in Ostend with these illegals who try to get onto ships is very damaging for customers. These transit migrants try and get into containers, and don't care whether there are goods inside them or not, or whether the cargo is perishable or not. What it boils down to is that a great deal of material damage is caused to goods. They also have to relieve themselves in that container, don't they? That causes a great deal of material damage and is no good to anyone. So the reputation of your port is closely linked to the security of your port.' (Interview Harbour Master).*

Above all, a port appears to be an *economic* reality. The port authorities are themselves responsible for paying for additional physical security measures (fencing, etc.) as well as the extra costs in terms of deployment of staff and resources. Also, carrying out checks leads to delays, which in turn means financial loss. It is self-evident that the security logic is on shaky ground with these economic laws. In relation to the phenomenon of transit migration, the presence of illegal migrants passing through is mainly seen by the port in terms of damage and extra costs. Not only does the damage they can cause to cargos cost money, the delays caused by checks are damaging to profits. In short, from an economic viewpoint it is crucial that the flows continue to flow (Bakker, 2007).

Welfare organisations of the cities of Ostend and Zeebrugge

> '"*We do what we have to do, from our social mission,*" so says the director of the local Centre for Social Welfare. "*In the organic law of the Centre for Social Welfare, this right is also enshrined in the first article, which stipulates that everyone is entitled to a human existence. It comes down to actually enforcing those rights.*"'

Both the Centre for Social Welfare and Outreach Work (*Straathoekwerk*) start from 'humanitarian' needs. This involves these bodies developing low-threshold, aid to transit migrants as a new target group. In contrast to the aforementioned actors, whose concerns generally involve protecting against the irregularities caused by the arrival of transit migrants within their own operations, the starting point for a welfare-based intervention is the needs of its clients. For example, by working proactively, social workers out in the field (such as outreach workers) try to build up contacts with the target group. On the one hand, the Advice Centre for Immigrants (*Adviescentrum voor Allochtonen*, ACA) provides legal advice relating to residency legislation and asylum procedures. On the other, migrants can use it for psychosocial help relating to financial and material problems, migration and the like. Voluntary repatriation is also organised in collaboration with the International Organisation for Migration (IOM). In addition, the ACA also supports families in collaboration with Refugee Work Flanders (*Vluchtelingenwerk Vlaanderen*, a non-profit organisation) and Caritas International.

Respondents in the social sector in both cities share their concern in relation to the local authorities. From their perspective the local authorities fail to address the issue of transit migration, for instance in Ostend, as no measures are taking to deal with the challenges this poses to the migrants & the residents of the city of Ostend.

> 'The main absentee is the government. Ostend is incredibly afraid of that knock-on effect, except where it involves the elderly and tourists and the culture-minded public. But they will not see that if you want to assume such airs; that also brings other aspects with it. What they do is close their eyes for as long as possible to the problems that arise and when nothing else can be done, panic breaks out. A local authority should take a more prominent role. ... But they are afraid of the knock-on effect if we were to develop the relief facilities.' (Source: Centre for Social Welfare directorate).

Local authorities of Ostend and Zeebrugge

The seaports of Zeebrugge and Ostend are also places that can be seen as *maritime nodes*. The city of Ostend also functions as a tourist attraction, which has major implications for objective, but above all, subjective security aspects. Second homes are regularly broken into, and the 'overload' created by this target group in the eyes of Ostend's residents results in waning support for all kinds of relief and aid initiatives. The overload is also damaging to the tourist image that the spa towns want to cultivate. This is where the tension between surveillance and tourism comes to the surface.

After finishing this empirical work, the local authorities in Ostend took initiatives in 2011 to deal with this issue which had been addressed by another policy-oriented research (De Ruyver, 2009). On the orders of the mayor of Ostend and the chief constable, police officers have to continue 'the hunt' for people with an illegal status in the city on a daily base, and increase the amount of 'stop and searches' of suspicious persons. In practice, this repressive policy implies that every day 20 people have to be arrested for a period of 12 hours in order to discourage their attempts to make a clandestine transfer to the UK, on the one hand, and to decrease the negative consequences of crime and disruption in the touristic heart of the city, on the other hand. However, a lot of critique has been formulated by human rights organisation since the approach does not affect the underlying social problems.[9]

9. http://www.hln.be/hln/nl/957/Belgie/article/detail/1315367/2011/09/07/Oostende-wil-illegalen-massaal-arresteren-als-ontrading.dhtml

Network analysis

Needless to say the logics mentioned above generate a tension in the relationships between the actors, the 'nodes' in the field who are responsible for dealing with transit migrants in Ostend and Zeebrugge. The research revealed that there is no form of network between the aforementioned actors. The exception is the working group on 'illegals passing through', which was set up in 2001 with the aim of developing informal joint ventures and promoting contacts between police services and, among others, the welfare sector. This revealed, among other things, that it was unclear where the tasks of one actor stops and where the responsibility of the other actor begins. Furthermore, the working group has only met once. In addition to mentioning this seed for cooperation, reference was made to a couple of related areas of tension that our empirical research revealed.

The first area of tension involves sticking to core tasks on the one hand and the need for commitment and goodwill on the other. The actors questioned in the field are aware that without partnerships they have insufficient strength to control the phenomenon of transit migration. This opens the door to the contribution of a number of welfare organisations that approach this phenomenon more from the perspective of providing assistance. Respondents within the various settings point to 'extra efforts' that must be made, assignments that have to be performed in addition to their 'traditional' set of tasks, but for which it is not immediately clear who actually has the mandate. This has a number of consequences. In practice, it seems on the one hand that by committing themselves to assignments that do not strictly legally form part of their mandate, the core tasks of various actors are seriously jeopardised. Not only does a large part of the Waterway Police's capacity now go towards handling migrants passing though, the local Centre for Social Welfare have in a certain sense become the victim of their own voluntarism, which significantly impedes their work with their own local deprived population. On the other hand, if each actor sticks to their strictly legal powers, there are a number of tasks that need to be done, but which do not get carried out. This shows that 'goodwill' helps build up a network with other nodes and formulates an answer to questions that arise when providing relief aid for transit migrants. After all, certain 'nodes' sticking firmly to core tasks can ultimately lead to them fighting each other, seen from the perspective of conflicting logics as outlined above.

A second area of tension linked to this is the tension between the often diverse (conflicting) sources of control to which actors are subject. A certain duality of function was identified, especially among the local police. On the one hand, reference is made to control (by citizens, judicial authorities, etc.), by which the democratic need to sometimes have to take repressive action is recognised. On the other hand, there is a certain understanding of the background and precarious position of people with an illegal status passing through. The result is that attempts are made to avoid an overly tough approach, so that among other things, welfare work can fulfil its role. The ambiguity in the way in which the police are assumed to act therefore largely concerns its legal missions in its capacity as supervisory and enforcing authority, which appear to get into hot water if this unilaterally controlling, repressive way of working goes too far. As a result, the performance of key missions by other bodies risks being impeded. After all, the fact that a number of other actors are organising themselves around this topic, starting from a different rationality, therefore means that each body recognises and respects the individuality of these other 'nodes' that are part of a 'nodal governance strategy'. In short, for the local police, it requires a constant balancing act in which they must follow up on the demands of administrative and judicial authorities without jeopardising the implementation of the vision, mission and values of other nodes within the broader network.

In other words, there are clear obstacles to developing a network between the key actors that are active in the surveillance of transit migration and the harbours that we studied. A network that currently does not exist.

4 Conclusion: some reflections on a possible nodal-network fallacy

In relation to the nodal-network fallacy described by Shearing & Johnston (2010), the above analysis suggests it is a useful perspective when analysing the surveillance of transit migration within the seaports of Ostend and Zeebrugge. It immediately becomes clear that the police are no longer the central security players in tackling the problems associated with transit migration. Several public, private, security and welfare organisations are positioning themselves in the approach to these problems. From a 'nodal governance' perspective, each of these actors can be described as a 'node' that can play a role in dealing with the diversity of defined problems in relation to transit migration (from problems for the immediate vicinity of the port, physical security problems within the port, etc.). Nodal mapping (analysis of the logic of each of these actors) revealed that different rationalities control the function of these organisations. The question immediately arises as to whether or not these actors form a network, since 'nodal governance' presupposes that a single central authority is no longer exercising control, but that a network approach is in place. The network analysis shows that the different rationalities stand in the way of harmonisation and optimum cooperation between actors. It seems as if some practical restrictions limit the ability to take the form of a network model. Different logics, but also conflicting goals (political, touristic, economical, etc.) inhibit further cooperation designed to organise surveillance around this new 'global' phenomenon on a local level.

In conclusion, it is clear that the phenomenon of transit migration in harbours can be considered to be an illustration of a phenomenon related to the globalisation of our network society with huge consequences for some locations. Flows of people, information and goods 'meet' each other in the harbours of Ostend and Zeebrugge and so do transit migrants. In practice these harbours are mainly defined as geographical nodes in which supervision and enforcement is needed to deal with the security issues related to transit migration. This view is present in the perception of the police, as main law enforcement agency working in the harbours. Still, the police are aware of the fact that larger mechanisms are influencing the effectiveness of their daily efforts. In their expression of 'battling a loosing fight' lies the awareness of mechanisms that they cannot address alone. This encourages them to explore partnerships with other actors in the field. Nevertheless, mechanisms beyond the network they can possibly build are at stake here. Push and pull factors related to policies in foreign countries are influencing the flows of transit migrants and the challenges they pose for locations such as the seaports of Ostend and Zeebrugge. Although the various kinds of local actors are experimenting in their networking, they find it hard to affect the push and pull factors that really matter. Although many of them are trying to think and act 'out-of-the-box', it is a hard job to make it work into practice. Often they are ultimately left with a lot of frustration and they return to their main logic (economic, social, security related, and so on). This logic, which has been functional for many years, can be considered to be a 'geographically limited' institutional logic that generates 'islands' that get challenged by the true nature of our society.[10]

As actions taking place within harbours are influencing and being influenced by actions taking place elsewhere, dealing with all aspects related to the phenomenon of transit migration has to be reconsidered. Nodal governance with its focus on nodal mapping and network analysis can be inspiring. It opens up discussion on the governing practices that deal with the true nature of recent phenomena (and the nodes, flows and processes they contain) and the existence of 'jungles' in which transit migrants end up living in nowadays (Nicholas, 2009). A more profound study of the phenomena of transit migration as such, in combination with a search for the key national and international actors and nodes, that can support a global approach – of which surveillance is only one aspect – is definitively a challenge for the future.

10. By the time of the publication of the text, the Ostend authorities stiffened their policy to prevent the flow of transit migration. The fences were replaced by higher ones, they changed the camera system with smarter camera's... .

References

Bakker, H. & Miltenburg, P. (2007). '''Nodaal'+'Nodal'= Denk Nodaal, Werk Lokaal.'' *Het Tijdschrift Voor De Politie*, 9, 24-28.

Bakker, H. P. S. (2006). 'Politie In Stromenland: Over Nodes En Netwerken.' *Het Tijdschrift Voor De Politie*, 10, 19-23.

Bekkers, V., Van Sluis, A. & Siep, P. (2006). De nodale oriëntatie van de Nederlandse Politie: over criminaliteitsbestrijding in de netwerksamenleving. Bouwstenen voor een beleidstheorie, *Erasmus Universiteit Rotterdam*. 119 p.

Bekkers, V. & Van Sluis, A. (2009). 'Nodale Politie In Nederland. Een Tussenstand Van Zaken.'*Panopticon, 2*, 49-54.

Boutellier, H. (2007). *Nodale Orde: Veiligheid En Burgerschap In Een Netwerksamenleving*. Inaugurale Rede Bij De Aanvaarding Van De Frans Denkers Leerstoel Veiligheid & Burgerschap Vrije Universiteit Amsterdam.

Burris, S., Drahos, P. & Shearing, C. (2005). 'Nodal Governance', *Australian Journal Of Legal Philosophy, 30*, 30-58.

Castells, M. (1996). *The Rise Of The Network Society*. The Information Age: Economy, Society And Culture, Blackwell Publishing.

Cyrus, N. & Vogel, D. (2002). Managing Access To The German Labour Market – How Polish (Im)Migrants Relate To German Opportunities And Restrictions. *Third Iapasis Report*. Oldenburg: University Of Oldenburg.

De Bie, S. (2005). *Illegale Migratie Met Als Bestemmingsland Het Verenigd Koninkrijk*. Unpublished thesis, Universiteit Gent.

De Ruyver, B. (2009), *Voor Een Veilig En Leefbaar Oostende. Een Onderzoek Naar De Dreigingen, Uitdagingen En Kansen*. Rapport In Opdracht Van Stad Oostende. 144 (Http://Www.Oostende.Be/Nieuws/667/Default.Aspx?_Vs=1_N&Id=2911) [02/03/2010]

Easton, M. & Gilleir, F. (2010). *Transit Migrants In The Harbour Of Ostend And Zeebrugge. An Exploration Of The Phenomenon From A 'Nodal Governance*. Unpublished Report.

Ferweda, H., Van Bolhuis, V. & van der Torre, E. (2008). *'Politie Op De Infrastructuur. De Nodale Oriëntatie In De Praktijk Gevolgd.* Advies- En Onderzoeksgroep Beke & COT Instituut Voor Veiligheids- En Crisismanagement.

Gilleir, F., Easton M., Ponsaers, P. & Cools, M. (2008). Nodale Politie In De Antwerpse Haven., Unpublished research report.Heymans, H. (2008). *Onregelmatige Vreemdelingen Onderschept Aan De Grens Of In Het Binnenland, Eerste Trimester 2008*. Intern Rapport, Sa Dga/Dac-Imm.

Heymans, H. (2009). *Onregelmatige Vreemdelingen Onderschept Aan De Grens Of In Het Binnenland, Derde Trimester 2009*. Intern Rapport, Sa Dga/Dac-Imm.

Johnston, L. & Shearing, C. (2003). *Governing Security: Explorations In Policing And Justice*. London: Routledge.

Jordan, B. & Düvell, F. (2002). *Irregular Migration: Dilemmas Of Transnational Mobility*. Cheltenham: Edward Elgar.

Nicolas, L. [Retrieved August 42009*. De Jungle Van Calais, (Http://Www.Mo.Be/Index.Php?Id=63&Tx_Uwnews_Pi2%5bart_Id%5d=25823&Chash=3431bb6bf3) [Retrieved March 25 02010]

Shearing, C. & Johnston, L. (2010). 'Nodal Wars And Network Fallacies, A Genealogical Analysis Of Global Insecurities', *Theoretical Criminology, 14*, 495-514.

Shearing, C. & Wood, J. (2003). Nodal Governance, Democracy, And The New Denizens, *Journal Of Law & Society, 30*(3), 400-419.

Tol, B. V. (2007). 'Nodale Oriëntatie In Relatie Tot De Controlefunctie Van De Politie: Niets Nieuws Onder De Zon In Stromenland?' *Het Tijdschrift Voor De Politie, 3*, 26-29.

Van Houtum, H. & Pijpers, R., Towards A Gated Community [Www] (Http://Www.Eurozine.Com/Articles/2005-01-12-Houtumpijpers-En.Html) [Retrieved March 25 2010]

Van Meeteren, M., Van San, M. & Engbersen G. (2007). *Irreguliere Immigranten In België; inbedding, Uitsluiting En Criminaliteit*, Acco, Leuven.

Vanoverschelde, J. (2007). *Stageverslag Scheepvaartpolitie Antwerpen.* Unpublished thesis Gent.

Van Poucke, W. (2004). *De Aanpak Van Illegalen Op Doorreis.* Unpublished thesis. Visietekst: Politie In Ontwikkeling, Visie Op De Politiefunctie, Rhc/Npi, Den Haag, 2005.

Wood, J. (2006). 'Research And Innovation In The Field Of Security: A Nodal Governance View'. In: in: Wood, J. & Dupont, B. (2006)., *Democracy, Society and the Governance of Security*, Cambridge, Cambridge University Press.

Policing flows and nodes: a Dutch interpretation

*Peter Marks and Arie van Sluis**

1 Introduction

In situations where many flows and nodes come together, such as airports and seaports, traditional methods of crime control and prevention are no longer suitable (Ruessink & Tops, 2006) and different roles for police are needed. This premise was adopted in the concept of nodal orientation in a strategy document called Police in Evolution (PIE) published by the Dutch Board of Chief Commissioners (*Raad van Hoofdcommissarissen*) in 2005. The strategy proposes that the Dutch police should focus on flows, i.e. on infrastructure such as road systems, waterways and communication networks, and on nodes, i.e. places where the flows of people, goods, money and information coincide.

The choice for nodal orientation must be seen in view of the social and technological developments that have contributed to a new type of criminal behaviour (PIE, 2005). Society has changed drastically: borders have been removed, mobility has increased and anonymity has increased. These typify a modern society that is gradually evolving towards a network society in which people, goods, money and above all information flows are becoming more and more important (Castells, 2000). Meeting places are no longer only physical; due to information technology they are also virtual, or a combination of both. The international capital market is a physical market on the floor of the Amsterdam stock exchange, but at the same time it has globalised and now offers knowledge and information about share prices via ICT streams. ICT and other technological developments enable the anonymity of individual movements in the 'land of flows' to be abolished. In nodal orientation new technologies are being applied in order to be able to follow people and goods.

Automatic Number Plate Recognition (ANPR) related activities are predominant amongst practices that can be labelled nodal (Van Bruul, Damkat, Van Dijk, Frenken, & Haasteren, 2008; Van Sluis, Marks, & Bekkers, 2011). ANPR is a technology that uses cameras to automatically register car number plates and compares the results with a reference database of number plates that have drawn the atten-

* Peter Marks and Arie van Sluis are Assistant Professors of Public Administration at the Faculty of Social Sciences, Erasmus University Rotterdam, the Netherlands.

tion of the police (Verhofstad, 2008). Nearly all Dutch police forces are using ANPR technology in one way or another. However, nodal policing is more than applying ANPR technology to traffic flows.

In this chapter we investigate the consequences for the Dutch police monitoring and operating major infrastructural flows and nodes, i.e. the Rotterdam seaport and Schiphol International Airport. We are particularly interested in the interpretation of the concept of nodal policing by the Dutch police, and the consequences this interpretation has on policing. That is, we are interested in how it affects the position of the police in security networks, how information is shared with and between the police and security network partners, what nodal strategies the police and network partners use, and the relationship between the security network partners and citizens. Our research question therefore is: How is nodal policing applied to infrastructural flows and nodes and what consequences does this have for the role of the police?

This chapter is partly based on a research report by Bekkers, Van Sluis and Siep (2006) commissioned by The Police Science and Research Programme. This programme funds independent academic police research. The goal of the research was to see what the concept of nodal policing means exactly, what its possible applicabilitions may be (both within the police force, but also in other domains), and what conditions need to be met to implement nodal orientation. For this research a literature study was done, followed by case scanning. Twenty key respondents, including members of the police force and academics, were interviewed. For this chapter we have updated the material with additional literature research and the findings of recent studies into Dutch nodal policing to get a general idea what typical Dutch nodal orientation is (Bruul, et al., 2008; Hoogewooning & Van Dijk, 2008; Van Dijk, 2008; Van Sluis, et al., 2011; Verhofstad, 2008).

We opted to focus on two major infrastructural flows and nodes because all possible aspects of nodal policing are contained in them. The description presented in the Section three is based on empirical work that was carried out in the Port of Rotterdam and Schiphol Airport. Public sources were used to compile the empirical account. Besides data from documents, semi-structured interviews were conducted with key respondents. An important selection criterion was that the respondents had key positions within the safety and security networks, i.e. directors of the security region Rotterdam-Rijnmond, a supervisor, a senior policy advisor and the commander of the Seaport Police, senior security and safety officers at Schiphol Airport, and a senior manager at Customs in the Port of Rotterdam. During all interviews we used a checklist that contained all elements of the typical Dutch interpretation of nodal policing as presented in Section 2. During the interviews the respondents were asked to elaborate on their opinions about nodal policing and the possibilities it could create or the dangers it may incorporate. The authors have translated all excerpts from the interviews that appear in the next section. In addition to the interviews, the authors observed the control rooms of the Seaport Police, the regional police and the harbour master. We also went into the harbour on boat to see how monitoring works on quaysides and wharfs.

In the next section, we will introduce Dutch nodal policing, comprising five key elements: the Dutch interpretation of the nodal orientation, nodal governance, information, nodal strategies, and privacy. In Section 3, we will present the two multi-node and flow cases, i.e. the Port of Rotterdam and Airport Schiphol. We will present the data in the format of the five key elements containing both description and assessment. In Section 4, we will discuss possible consequences when applying the nodal orientation. Finally, our conclusions are to be found in Section 5.

2 Dutch nodal policing

Shearing refers to nodal policing in the context of the growing diversity and multiplication of providers of policing. 'The police are only one node in a network that works to govern security' (Shearing, 2005, p. 58). Nodal security networks are important weapons against new, network-like manifestations of crime in the network society. In these networks the police are not by definition in charge, and since information is fragmented, the police have to coordinate and cooperate with other security partners (Boutellier, 2007). In a complex and fragmented system of supervisors, law enforcers, inspectors and guards, it is no longer obvious that the police are the central symbol of protection.

- *Nodal policing: The Dutch interpretation*

The Dutch interpretation of nodal orientation combines a perspective that considers the police a player in security networks with one on policing in infrastructural networks. Nodes and flows have become important bearers of economic and socio-cultural processes and activities. There are five distinct types of flows: people (using physical space), goods (using physical space), energy (mainly using physical distribution networks), capital (using ICT-infrastructure in the virtual worldwide capital market as well as physical space), information and communication (both using virtual space). These flows make use of physical distribution networks and occupy space in the virtual world. Linking of flows and infrastructure creates crucial new vulnerabilities at physical nodes, where the flows come together.

The vulnerability of nodes and the vital role that they play in the network society offer interesting loopholes for criminal organisations and terrorist groups. Criminals and terrorists understand the possibilities that this context creates and have not hesitated to make use of them, as noted in PIE (2005, pp. 21-22). Dutch nodal policing involves the intensive control and monitoring of infrastructures and flows of people, goods and money that move along these infrastructures. In order to limit mobility of crime and support timely identification, structural checks at network nodes are necessary, thus removing anonymity and invisibility (PIE, 2005, pp. 8-79). The various authorities need to consider many factors (simultaneously). It is not only safety and security that directs actions, but also fiscal aspects, transport, trade, labour organisation and so on that have to be taken into consideration.

- *Nodal security governance*

Infrastructural nodes and flows have many different 'owners', public as well as private, each of them with their own interests, duties, responsibilities, and powers in supervision, enforcement, and investigation into suspect activities. The nodal orientation, i.e. monitoring flows and nodes, and nodal security governance, i.e. mutual cooperation between parties in order to determine security risks and criminal behaviour and to take effective measures to improve security and to reduce crime, reinforce each other (Van Sluis & Bekkers, 2009).

Police organisations, supervisors, inspectors and others who are responsible for maintaining law and order should work in a problem-oriented manner (Sparrow, 2000). That is, the key duty of a company or organisation is to tackle problems within a network of organisations capable of contributing to address specific problems. Cooperation- and nodal security governance arrangements (horizontal arrangements) become relevant for steering and control. The number and variety of the actors involved, public as well as private, locally as well as nationally and internationally, require this (Bekkers & Van Sluis, 2009, pp. 51-52). A cooperative attitude from every supervisory authority requires a willingness to share information. There must also be a basic level of mutual respect for one another's work if all parties are to attain common goals. Per flow, node or infrastructure, the question of where the pre-eminence of investigation lies and what the consequences are for the police's role become relevant. This requires a mutual defining and solving of the problem as well as cooperation with civilians and other institutes. However, at the flow level it is hard to make connections with civilians.

- *Information*

In order to build up a strategic position, there must be an exchange of information between parties involved. Security partners need to collect and combine both hard and soft information and develop adequate interventions. The quality and availability of information from various sources is crucial. By linking the databases of the various partners, vulnerabilities and risks plus opportunities and possibilities within the security chain can be exposed. Information has to be filtered, interpreted and linked within a context, for instance, by using risk definitions, selection and interpretation based on profiling. This requires not just static information, but also dynamic information, i.e. actual information about what is going on in real time.

'Intelligence' is created by interpreting existing information and connecting them in a certain way. Intelligence also refers to the ability to consider the information in combination with existing knowledge and experience in context. Information from database exchange provides risk profiles that can be used for intelligence-led enforcement and criminal investigations (Van Sluis & Bekkers, 2009, p. 88). The quality of risk definition, analysis and evaluation are paramount for risk profiling. It takes knowledge and skills to see certain patterns, to recognise them, but also to put them into perspective (Ottinger, 2010). Statistical connections do not necessarily mean there is an actual connection. It is also important to be able to interpret these patterns by paying attention to the specific context in which a pattern does or does not take place. The appropriate use of information fits into the longer-term development of its systematic use and exploitation and an improvement in the information warehouse (Hoogenboom, 2009, p. 66).

- *Nodal strategies*

Each flow, infrastructure and node lend themselves to specific applications for interventions that suit their characteristics. In general, the police are able to apply two types of interventions: setting virtual traps at the access or exit points of flows, and moving with the flows within a specific type of infrastructure. Examples of the first type of intervention are traffic controls at the access points of highways, at the servers of internet providers, and the scanning of ships for the presence of nuclear material with radio frequency identification. Examples of the second type (exit points) are traffic patrols on highways, taking part in discussions in an extremist or religious fundamentalist chat room, and registering transactions that involve large sums of money in the financial flow. Applying either or both strategies usually involves the use of high tech detection technology that is linked to the characteristics of the flows and nodes. Operational follow-up after a technological hit is essential for the effectiveness of the process (Bekkers et al., 2006, p. 83).

Technology simplifies the identification of not only the turnover but also the status of goods, persons and other movements by applying a combination of monitoring, detection and identification techniques, whether or not in combination with biometrics. 'Police in Evolution' refers to the so-called catch-ken or catch-scan techniques where, for instance, scanning licence plates of cars passing by is automatically linked to information in various registers, such as police registers, licence plate registration and municipal administration bases. This way information regarding the status of vehicles or people can be obtained quickly.

- *Privacy*

Intensified monitoring of flows and nodes can pose a threat to the privacy of citizens because non-suspects will also be monitored and surveyed simply because they are moving within a particular flow. The authors of the PIE report believe that citizens would accept this intrusion of their privacy, and refer to a 'social contract' in which citizens accept that the expansion of some police powers enhances their ability to move around freely and safely. There will always be tension between protection by the government and protection from the government; the right balance is not static but shifts and adapts along with the social context (Welten, 2006).

According to PIE, the powers and duties of law enforcement agencies should be determined in close interaction with administrative authorities and citizens because this increases the legitimacy of the law enforcement agencies in solving social problems. However, the position of the police in a flow or node differs from their position in the real world, in area-bounded policing. Whereas community and area-bound policing are connected with existing political administrative structures and the habitat of citizens, there are no identifiable citizen communities in flows and nodes (ZHP, 2009). Successfully securing nodes and flows within the local boundaries has certain consequences for security partners and their relationships.

Linking the various existing sources of information creates new information power and intelligence about the whereabouts of people, thus opening doors for new and more refined forms of control (Bannister, 2005). In this context, it is important to note that nodal policing is not just a technology-driven intervention. To prevent the concentration of information from being abused, a system of checks and balances must be guaranteed. In this case, who inspects the inspector? And are we in charge of the technology we plan to use (De Mul, Müller & Nusselder, 2002)? People moving within particular flows are monitored even though they are non-suspects. The idea of the police watching impact negatively on society's support for the police (PIE, 2005, p. 76) A more detailed strategy for dealing with aggregated information and the related privacy implications should be realised (Van Sluis & Bekkers, 2009, pp. 90-91).

3 Cases

The Port of Rotterdam: multi-nodes and flows

The Port of Rotterdam is located on the North Sea and is fully accessible 24 hours a day, 7 days a week. Many different companies are active in the port, varying from storage and transhipment to industrial processing. In addition, the port's European hinterland, with over 500 million consumers, is vast and easily accessible. The port has five different modes of transportation, i.e. train, inland shipping, roads, maritime transport and pipelines (Port of Rotterdam, 2009). The port is an extremely complex and dynamic multi-nodal and multi-infrastructural intersection, with a global reach. Flows merge together in nodal points of the Port of Rotterdam; these in turn are connected to other large economic regions and form intersections where monitoring and surveillance come together.

- *Governance*

To create a safer and securer port the most important security actors in the port of Rotterdam, i.e. the Seaport Police, Customs and the Port Authorities, have defined five different flows and/or nodes, i.e. inland shipping, cargo, seagoing vessels, port security and border control. In these flows a variety of supervisory agents have to work together in different constellations, in varying degrees of diversity and intensity, and in an integrated approach (Transport and Water Management Inspectorate, 2009). This approach fits the Dutch Inspection Reform Program (Renewal Enforcement, 2009) that aims for more effectiveness in enforcement, with less inconvenience for the business sector. Supervisory agencies cooperate more closely by means of better information exchange, coordinated inspection programmes, joint risk analyses, alert enforcement and transfer of duties. Enforcement is focused primarily on high-risk activities and companies, grounded on information from and cooperation with the business sector: companies with a good record are subject to fewer checks. This is called 'horizontal enforcement' (Renewal Enforcement, 2009). Mutual adjustment in advance, based on cooperation, is more effective than retrospective control.

The logic of area-bounded policing does not stretch to the port area since there is only one citizen. The most important residents are industries, trade and commerce. The area is a mixed one, with public and private domains. Traditionally, the Port of Rotterdam is dominated by powerful private interests groups, such as the Port of Rotterdam Authority, that consider the port as their own backyard, where they do their business out of the public spotlight. It is difficult for the agencies that represent public interests, such as safety and security, to get access to the fixed and closed networks that exist in the port. Security and safety have to compete with economic interests, which are often considered to be more important. According to an interviewee, 'Authorities and actors within the port must recognise the need to balance economic and security interests. To do so, they need a clear and unbiased perspective on port-related security issues.' Improvement of security levels leads to image improvement and thus to a better competitive position of the Rotterdam port. Or as an interviewee put it: 'Security sells.'

In recent years, many elements of supervision and law enforcement have been placed within the tasks and responsibilities of a variety of both public and private actors within the port. At this very moment, this process is visible in the transfer of enforcement related police tasks to the local city police (*stadstoezicht*) and to the Harbour Master Division. The Seaport Police consider a nodal orientation to be indispensable given the complexity and vulnerability of the port area with the presence of many players in the enforcement field of security, conflicting interests, vital infrastructure, and a strong focus on economic and environmental aspects (ZHP, 2009). The Seaport Police aim at creating a secure and safe port by being a professional, reliable and leading network partner. They focus on the totality of flows: people, goods, shipping locations and destinations and the com-

plete logistic chain, i.e. on transfer points between sea and inland shipping, and between inland shipping and vehicle or railway traffic.
Among the multitude of supervisory agencies and the complexity of interests, the Seaport Police do not have the same obvious position and authority as the police have in area-bounded policing. However, the Seaport Police do claim a directing role in the chain of enforcement, control and investigation, to maintain a natural and automatic power, whether it is to embark and inspect ships or to confront people committing violations.

- *Information*

The Seaport Police are initiating, developing, participating and advising in numerous technological and administrative developments and partnerships within the port, i.e. Expert Centre Harbour (ECH), AQUAPOL, Maasvlakte 2, Mainport Development and the International Ship and Port Security (ISPS) Code. Information warehouses are being constructed to ensure the access to vital information in case of emergencies. In an interview, the director of the safety region clarified:

> 'What we do here is build virtual organisations, organise the network and facilitate all this through an information warehouse. Which means that I have a variety of co-thinkers and a lot of information. Hence I have made a connection between parts of the chain and a connection within the fragmented chain. Everybody uses one another's shared information and I keep it stored in a central place.'

ECH is a good example of how information is exchanged from a nodal perspective. ECH is administrated by the Port Consultation, in which all the ECH organisations are represented at a strategic level. All participants within the ECH (i.e. Seaport Police, the Royal Marechaussee[1], Customs, the National Crime Investigation Service, the Fiscal Information and Investigation Service/Economic Investigation Service (FIOD/ECD) and the Social Intelligence and Investigation Service (SIOD)) use the same Basic Provision Investigation system (*Basisvoorziening Opsporing*) to store information and data. The accessibility of national and international information systems enables ECH to carry out analyses for weigh reports and the development of project proposals. Before the Port Consultation takes a decision, the owner of the relevant information has to agree to the use of that information (De Gier & Helsdingen, 2009). On the whole, ECH has proved to facilitate communication and the exchange of information between agencies, and has boosted the growth of mutual trust, to enable better capacity management and to prevent activities from overlapping (Poortman, 2009). However, the exchange and sharing of information between law enforcement agencies runs into difficulties because of legal restrictions that are applied to safeguard the integrity of government. Currently, possible solutions are being investigated for the identi-

1. They are a kind of military police constabulary.

fied problems. Despite all this, information flows as freely as possible between ECH participants because mutual trust has grown due to the very short and informal lines of communication (Poortman, 2009).

- *Nodal strategies*

Since 2002, the Seaport Police have been using the Expandable Integrated Maritime Shipping System (*Zeescheepvaart Uitbreidbaar Integraal Systeem* – ZUIS). ZUIS is a unique system. All relevant information about seagoing vessels, countries of departure and destination, ship locations and crewmembers is entered into this system. ZUIS is also linked with other databases, including police databases, for example, those with information on wanted or reported persons. The system operates with fixed parameters to create a risk assessment and to determine the appropriate type of check for the vessel. The great strength of the ZUIS system is its highly flexible application possibilities. Risk indicators can be adapted immediately based on accumulated intelligence, which enables the prompt generation of new risk indicators combined with the appropriate operational deployment.

Customs has developed risk profiles to select potential risky shiploads, using an electronic system called Prisma. All shipping companies are required to electronically deliver shipload information. Every shipload gets a risk profile based on a combination of the type of goods, the shipping company, the country of origin, the country of destination, and the route. Prisma automatically generates hits when a risky shipload is indicated. These indicators are laid down in the Customs Act. However, a trained eye will always be necessary in addition to hits generated by Prisma.

As of May 2009, all bridges and tunnels that give access to the Botlek-Maasvlakte are constantly monitored with ANPR cameras. On a regular basis joint actions on ANPR findings are held in which numerous supervisory agencies participate and in which a variety of databases are linked.

The Port of Rotterdam is well known for its high standards of nautical services, the safe and smooth handling of shipping and a Vessel Traffic Guidance System of high quality. The Division Harbour Master contributes a great deal to these high standards with its 570 specialists, 10 patrol vessels and 100% radar coverage (Harbour Master 2009). All along the river and docks there are radars that make it possible to follow every ship in and around the port, also when they are moored or still anchored outside in open seas. Ship movements and positions are depicted using the dynamic port chart. All ships get labelled based on the cargo of the ship, so one can distinguish ships with petrochemical materials or fuel, or 'normal' and safe products. However, as an interviewee clarified: 'to date only seagoing vessels are depicted, because for inland going vessels they need more information and also too many dots for every ship will appear on the dynamic chart, making them undistinguishable.' All these activities are in coordination and collaboration with those responsible for safety in the area who are situated in the same building.

With respect to camera surveillance, the Seaport Police work closely with the Port of Rotterdam Authority, for example, regarding the radar camera. All trajectories that involve regular camera enforcement are clustered in the ECH. Seven different disciplines are brought together, with their own backgrounds and their own information. Information is exchanged based on protocols and on the agreements in the Dutch Inspection Reform Program. The Seaport Police monitor the follow-up of all signals, for example, with regard to inland shipping, whilst other agencies have a coordinating role. In these cases, the Seaport Police draw up reports of their findings.
Every vessel that anchors in Rotterdam is checked. There is a stop at the Parkhaven where vessels bunker fresh supplies. However, crew can disembark here without being seen, which is a weakness in terms of the Schengen treaty. Cameras are being installed and the images will be checked.

Customs moves with the flows of goods by monitoring ship movement, and it also checks the access to the port with container scans. Following a hit in Prisma or when annalists have doubts, based on intuition or experience, that the actual content of a cargo may not match the one registered, a ship is scanned with X-ray equipment. Customs can deploy fixed scans, mobile scans and drive-through scans for trucks. Scans may be followed by a physical inspection. Annually 40,000 to 50,000 containers are scanned, which is about 2% of the non-EU flow. The percentages of hits and the success ratios are not made public. However, statistics on captures and hits are shared with other custom organisations in the EU, through established channels, based on treaties. Forty nuclear detection gates check all transport that enters or exits the port for the presence of nuclear material. Logistics with regard to scanning are dealt with after preliminary consultation with forwarders, shipping companies and other parties involved.

- *Privacy*

In countries such as the US and UK, the separation between inspection and detection has been lifted due to the heightened political and social priority that is assigned to security. In the UK, for instance, customs is permitted to link information regarding movement of persons to the goods flow. This was in place before September 11. However, in the Netherlands customs are limited when it comes to exchanging information on goods. This impedes collaboration with custom authorities in other countries. After all, customs exchanges information with custom authorities in other countries. The Port of Rotterdam is a partner in CSI, Container Security Initiative. Shipment information is exchanged with the USA. Rotterdam is a 'second line of defence'. Shipments for the US are inspected in Rotterdam. However, due to legal limitations and the separation of inspection and detection, custom authorities are not permitted to exchange information with custom authorities regarding persons.

Schiphol: nodal policing at the Dutch national airport

Amsterdam Airport Schiphol is Europe's fifth largest airport in terms of passengers and third largest in terms of cargo. Amsterdam Airport Schiphol is an important contributor to the Dutch economy. It serves as one of the home bases for Air France-KLM and its SkyTeam partners.
Schiphol is a node that connects various types of infrastructure, not only air, but also road and rail; there are flows of people and goods. Schiphol is also a node for the activities of a wide range of private and public organisations. The leading strategic idea is that the airport should be a perfect stopover during travel. This is the concept of the 'Airport City'; a dynamic place where people and companies, logistics and shops, information and entertainment meet and support one another (Schiphol Group, 2010).
Obviously, security is an important issue at Schiphol. The fear of terrorist attacks triggered increased attention to security. The armed diamond robbery in February 2005 was a turning point as well, according to the senior Security Officer Schiphol Group. The robbery took place on Schiphol terrain. The stolen car was able to freely leave the area, which led to a great deal public amazement and anger. The diamond robbery gave the Dutch parliament cause for an investigation and prompted the Dutch government to establish the Committee Entrance Control Schiphol (the so-called Committee-Oord, named after its chairman). A start was made with a combined and coordinated public-private approach to security and crime.

- *Safety and Security*

At Schiphol, security and the actors involved have many faces. Security mobilises a variety of public and private parties with different tasks, powers and responsibilities in the field of inspection, maintenance and investigation. This brings about the need for structured cooperation between the parties and an adequate political-administrative embedding. A distinction can be made between safety and security
Safety refers to air safety, safety of working conditions, environmental safety, traffic safety and fire safety. The common denominator is protection against unintentionally dangerous situations. Security refers to protection against intentional and deliberate criminal acts. Since changes to aviation law in 2003, Schiphol is legally responsible for the execution of preventive security tasks at the airport, but under government supervision (Schiphol Group, 2010). Each year hundreds of millions of Euros are invested in all kinds of measures due to 9/11. The costs of these measures are paid through airport taxes imposed on airline companies, which they in turn claim from passengers by way of an extra security charge on tickets. Private security companies, mainly Amsterdam Airport Schiphol, airlines and airfreight companies, perform security activities. The Royal Marechaussee make sure that the law is enforced and supervise the execution of

the security measures by the private sector. The role of the regular police is a modest one. The National Coordinator of Terrorism prevention (NCTp) is, among other things, responsible for adequate security of civil aviation in the Netherlands and is also responsible for consistently assessing security.

- *Governance*

In the Safety Platform Schiphol (SPS), all companies at Schiphol Airport collaborate on safety. This platform is an alliance of all companies that are involved in aviation, such as Dutch and foreign airlines, Airport Schiphol, Dutch Air Traffic Control, service providers, catering and cleaning companies and fuelling companies.

All sixteen inspectors at Schiphol work together in the Front Office Schiphol, which has been operational since 2008 and is part of the government Inspection Reform Programme. It is a coordination and information exchange to reduce the burden and improve the quality of inspection. The motto is: 'more effect, less burden'. There is one contact point for companies at Schiphol, making inspection clearer. The goal is that in the future companies will only have to supply their information once. In addition, mutual risk analyses are drawn up and an integrated inspection programme is created. The SPS is the contact point for the Front Office, in part because the safety risks occur on the common ground which the safety management systems of individual companies share (Schiphol, 2008).

Public and private parties collaborate in the Platform Security and Public Safety Schiphol (established on 25 January 2005) to handle safety and security together. A better information position, supported by technology and integrated activities, allows mutual risk analyses to be drawn up. Mutual risks in security and safety are then more manageable, duplications are prevented and parties can streamline their processes better. The platform removes the boundaries between businesses and government, and makes topics open to discussion. The chairmanship is shared between the National Coordinator of Terrorism prevention and the president of the Schiphol Group. This enables win-win situations, which in turn helps mutual initiatives to take off faster. Examples are the use of cameras and the shared control room. The costs are divided equally between government and Schiphol. The platform enables a coherent approach to criminal and security problems by introducing centralised management. The chairmanship of the NCTp has created quick access lines to 'The Hague' making it possible to mobilise political support if necessary.

Public-private collaboration and mutual trust between the actors involved in the industrial zone at Schiphol are important success factors, according to a senior security officer. However:

> 'In business areas you can get collaboration off the ground, but sometimes our meetings are a bit of a bun fight. The General Intelligence and Security Service (AIVD) sometimes refuses to share information because other parties are present that are not allowed to have this information. They are requested to leave the room. Next, the AIVD tells us: "Our assessment of threats hasn't changed". This is no reason to send them out of the room. If you want to collaborate, you have to trust others.' (Source: interview).

- *Information*

Initiatives by the platform also lead to an improved information position. All information from immigration, justice department, customs, law enforcement, marechaussee, armed forces and KLM are stored in a database. Based on authority and signed contracts everyone can retrieve the information they require. The platform offers a forum for difficult subjects, such as the problem of privacy in case of camera supervision and the quality of the exchange of information. However, besides mutual interests there are competing interests and legal barriers making the gathering and exchange of information difficult. Schiphol collects data for their own business processes and for better service to their customers. However, this information may also be valuable for government institutions.

> 'In most parking places in the Schiphol area, number plate recognition is being used. We want to advice our regular customers on where to find the cheapest parking rates. For this reason we need to identify them. At a certain point in time we shared this information with the Royal Marechaussee. Their database showed that 21 vehicles parked in our garage were actually stolen. Now, the Marechaussee want to connect our business management database with theirs, but there is no legal framework for this. This is under construction. Technically, it poses no problem whatsoever. Besides, it would be foolish not to do this. However, our prime target remains to serve our customers.' (Source: interview).

Schiphol does not pass on any passenger information to third parties. These data are the property of the airlines and the use of these data is subjected to strict regulation. The Marechaussee pass on information because they do passport control. Schiphol is primarily interested in goods, not people, and they consider checking people a government responsibility. The mutual sharing of information is in need of improvement, or as an interviewee of Schiphol stated:

> 'You should arrest terrorists before they arrive at Schiphol. This requires sharing of information. The General Intelligence and Security Service did not have any information on the subjects that caused such a threat in the London subway.'

The Data Protection Authorities are involved in this discussion.

- *Nodal strategies*

A leading principle in monitoring passenger flows is the strategic concept of 'self service airport'. Passengers enter through the front and leave through the back, via the 'boarding area'. Within this concept the nodal strategy aims at making deviant behaviour visible by focusing on entrees and exits of flows, for example, by setting up physical or virtual traps, like body or iris scans, or by making movements within a flow transparent through camera surveillance.

At Schiphol itself a network of cameras is operational. The cameras are used for mutual use of security footage. 'A network is being set up in which all parties involved can track, register and reconstruct their security processes via cameras. Public and private partners can each use visual materials for observation, identification and investigation, each for their own specific purposes and responsibilities.' (Schiphol Group, 2006). Schiphol would like to conduct more supervision using more cameras and less staff, by having 'smart cameras' that switch on if a person in the masses acts suspiciously or if other suspicious situations occur. In principle this would make it possible to work in a proactive manner.

Hand luggage goes through X-ray machines at the entrance. Checked baggage is automatically checked for explosive and dangerous materials and objects. This is done using a fully automatic system that has been in place in the EU since 2002. Previously suitcases were not checked. At present the baggage check is 100%, as has been the case worldwide since 2006.

Body scans are also used, they work with X-rays that check for swallowed material. The security scan is used for security and customs control of passengers. The security scans work by reflecting an image of the body contour. In this way it will be noticed if a passenger is carrying illegal objects on their body or if goods are being smuggled under their clothes. At present, passengers can decide whether they wish to go through the security scan. If they opt not to, they have to go through the usual checking procedure. In December 2008, the European Parliament objected to making this procedure compulsory due to the privacy aspects and possible health risks. The discussion about body scans has come up again as a result of the thwarted bomb attack on Christmas Day in 2009. During a flight from Schiphol to Detroit a young Nigerian man tried to set off a bomb which had gone through the detection gates unnoticed, but which would have been picked up in a body scan.

Most staff entrances are equipped with devices that can check the biometrical information on Schiphol passes. Staff passes include a digital image of the iris. Software is being developed that will enable suspicious persons to be detected faster and easier, thus making it possible to follow the person concerned sooner and more effectively. A 100% check is in place for staff members.

In addition to passengers and goods, Schiphol also monitors railways and roads. Tests have been conducted with the Dutch Railway Company using cameras on platforms under the terminal. That way unattended luggage can be detected. Also on the road in front of the departure hall a test was conducted with aimed camera inspection, which verified license plate numbers. For instance, a relatively high amount of stolen cars were detected. Criminals use parking lots to park stolen cars for the time being. As of July 1, 2008 all vehicles that enter the platform at Schiphol airport are subjected to a security check.

- *Privacy*

According to one of our interviewees, privacy has been made too much of an issue:.

> 'Privacy gets too much attention. You can put an end to the fight against terrorism if objections are raised against the use of cameras, because they produce the kind of information you don't get out of a computer.'

4 Discussion

The dangers associated with the use of high tech in nodal strategies are naïve instrumentalism and an unconditional trust in the possibilities and blessings of technology. Even if technology is used with the best intentions in the world, various kinds of unintended (desired and undesired) effects occur because technology has a dynamic all of its own, for example, following persons and/or monitoring their behaviour with cameras, by releasing information onto public and private databases and by interpreting existing databases that are connected or aggregated, affect personal privacy. Intelligent camera supervision can be a 'privacy enhancing technology' which has a positive effect on privacy, because images are only recorded if required based on detection of certain movements. But in general such a proactive use of technology has still scarcely begun. In present day nodal strategies, large groups of non-suspect persons are followed or have their anonymity removed and hence are labelled as 'suspicious'. To what extent can the freedom of large groups of civilians that move within certain flows and nodes be affected in an effort to offer more security? And to what extent does the construction of enormous databases with unknown benefits really contribute to more security? The critical question is whether there is a balance between the contribution of technological applications for tracing security threats and the rights and obligations for the protection of personal privacy.

The legal framework with regard to the position of the nodal practices and what to do with (stored) information is underdeveloped in the Netherlands. Information is exchanged with other services within the boundaries and limitations of the law. The issue is whether it is permitted to link the records and the stored information. Information may only be used for the purpose for which it is collected, for instance, to levy taxes or other fiscal matters. It is not intended for investigative purposes. This protection method has to do with the fear of misuse of information and the necessity to guarantee the integrity of government actions.

Only very recently a legal framework is being 'reconstructed', based on the recommendations of the committee Brouwer-Korf (2009) that advised the Dutch government on privacy issues in relation to security. The committee observed that the discussion about careful handling of personal data has been dominated by an almost exclusively legalistic approach. As a result, the legal system has given virtually no guidance to professionals about how to deal with personal data for security purposes and has left them ill-equipped to put the legal system for handling personal data into practice. In the Port of Rotterdam, privacy issues are pragmatically dealt with within the legislative ranges and possibilities, while at Schiphol Airport they seem to be thought of as a hindrance.

5 Conclusions

In the Port of Rotterdam and at Schiphol Airport certain nodal practices are slowly surfacing: one can say that the nodal orientation is slowly but surely developing. That is, public and private agencies have to find one another, for collaboration and cooperation, exchange of information and to lean on one another's expertise and intelligence. Several information warehouses are being constructed and the parties deliver their information that is accessible, for instance, in cases of emergency.

In the Port of Rotterdam, agreements and covenants are made for the exchange and usage of 'shared' information to tackle safety and security issues in several collaborations. This is done within the context of legal rules that are in their developing stages. Actors cooperate in nodal security arrangements relevant for steering and control, i.e. there some forms of nodal security governance. However, this is still in the starting stages, and although most parties have similar views on safety and security, actors still operate from their own perspective. At Schiphol Airport all safety and security partners are collaborating in safety platforms and coordinate their information exchange. However, the mutual sharing of information is still in need of improvement.

The Seaport Police would like a more leading position within the horizontal playing field. Vertical repositioning means emphasising the authoritative, professional and independent role of the police as well as creating an unambiguous and natural authority, similar to the natural operational authority and acceptance of the traditional policing. It does not mean withdrawal from cooperating within the horizontal playing field. On these matters, the Seaport Police have rethought their strategic vision, in which the nodal orientation is now fully embedded. At Schiphol Airport, actors other than the Royal Marechaussee are the leading actors in security and safety. Their role is a more modest one in the background, i.e. supervising the other parties.

All in all, the role and position of the police in the port of Rotterdam and Schiphol Airport are different because they have different tasks as well as operate in different security networks. In the multi-nodal port and airport, the police have to consider whether they want to be the leading actor or not, i.e. they have to re-think their strategic position. This requires tailor-made solutions and thinking about core tasks. Core tasks must be executed within multi-disciplinary processes and hybrid public-private structures, in which the borders between agencies tend to 'fade away' (Hoogenboom, 2009). This serves to complicate the execution of core tasks. Nodal policing also touches the police's relationships with citizens, especially with regard to individual privacy. It is not just a technology-driven intervention, but also politically sensitive and something that is not just for the police to decide upon. It is essential that it is clear who makes the decision about collecting personal data, who is responsible, how these decisions are made and how they are accounted for. That is, nodal policing requires effective democratic control.

References

Bannister, F. (2005). The panoptic state. Privacy, surveillance and the balance of risk. *Information Polity, 10*(1-2), 65-80.

Bekkers, V., Sluis, A. van & Siep, P. (2006). *De nodale oriëntatie van de Nederlandse politie: over criminaliteitsbestrijding in de netwerksamenleving. Bouwstenen voor een beleidstheorie.* Rotterdam: Center for Public Innovation.

Bekkers, V., Sluis, A. van (2009) Nodale politie in Nederland. Een tussenstand van zaken, *Panopticon* 2, 49-56

Boutellier, H. (2007). *Nodale orde: Veiligheid en burgerschap in een netwerksamenleving.* Inaugural speech for the acceptance of the Frans Denkers leerstoel Veiligheid & Burgerschap Vrije Universiteit Amsterdam.

Bruul, I. van, Damkat, P., Dijk, W. van, Frenken, W. & Haasteren, P. van (2008). *Leiderschap als knooppunt. Een studie naar de rol van leiderschap in de diffusie van nodale oriëntatie binnen de Nederlandse politie.* Warnsveld: Thesis School voor politieleiderschap.

Castells, M. (2000). *The rise of the network society. The information age: economy, society and culture.* Cambridge, MA; Oxford, UK: Blackwell Publishing.

Committee Brouwer-Korf (2009). *Gewoon doen, beschermen van veiligheid en persoonlijke levenssfeer.* Den Haag: Rapport van de Adviescommissie Veiligheid en persoonlijke levenssfeer.

De Gier, M. & Helsdingen, I. (2009). *Working apart together. Een onderzoek naar samenwerking.* Afstudeerscriptie Bestuurskunde: Erasmus Universiteit Rotterdam

De Mul, J., E. Müller & A. Nusselder, A. (2001). ICT de baas? Informatietechnologie en menselijke autonomie. In Zouridis, S., Frissen, P., Kroon, N., Mul, J. de & Van Wamelen, J. (Eds). Deel 7 van Internet en Openbaar Bestuur. Den Haag

De Mul, J., E. Müller & A. Nusselder (2001). ICT de baas? Informatietechnologie en menselijke autonomie, Onderzoeksprogramma Internet en Openbaar Bestuur, Den Haag.

Harbourmaster (2009). *Port and terminal information project.* www.harbourmaster.nl

Hoogenboom, B. (2009). Dingen veranderen en blijven gelijk, *Justitiële verkenningen, 35/1*, 63-77.

Hoogewoning, F.C. & Dijk, A.J. van (2008). Onlosmakelijk verknoopt: lokale en nodale oriëntatie. *Het Tijdschrift voor de Politie, 70*(4), 28-32.

Ottinger, G. (2010). Constructing Empowerment Through Interpretations of Environmental Surveillance Data. *Surveillance & Society 8*(2). 221-234

Poortman, S. (2009). *Samenwerking in de Rotterdamse Haven: een onderzoek naar de samenwerking tussen de Douane Rotterdam en de Zeehavenpolitie.* Afstudeerscriptie Bestuurskunde: Erasmus Universiteit Rotterdam.

Port of Rotterdam (2009). *Organisatie ZHP.* www.zeehavenpolitie.nl

PIE (2005). *The Police in Evolution; vision on policing.* The Hague: NPI.

Ruessink, H. & Tops, P. (2006). *Nodale Oriëntatie: In de knoop moet je zijn.* Report of the lector conference Nodal Orientation, November 10th 2006, Zwolle, Politieacademie.

Renewal Enforcement (2009). *Inspection Reform Program. More effect, less burden!* Den Haag: An Inspection Council Bureau publication.

Schiphol (2008) Eenduidig Toezicht Schiphol. Hoofddorp: Amsterdam Airport Schiphol.

Schiphol Group (2006). *Corporate responsibility in 2006 Amsterdam.* Amsterdam Airport Schiphol.

Schiphol Group (2010). *Schiphol innovatief vanaf het begin.* Amsterdam Airport Schiphol.

Shearing, C. (2005). Nodal security. *Police Quarterly, 8*(1), 57-63.

Sparrow, M. (2000). *The regulatory craft. Controlling Risks, Solving Problems, and Managing Compliance.* Washington: Brookings Institution Press.

Transport and Water Management Inspectorate (2009). *Water Transport Supervision Plan.* Den Haag: Koninklijke De Swart.

Van Dijk, A.J. (2008). Ruimte voor de politiefunctie. De lokale en de nodale oriëntatie. In Boutellier, H. & Steden, R. van (Eds). *Veiligheid en burgerschap in een netwerksamenleving.* Den Haag: Boom Uitgevers.

Van Sluis, A., & Bekkers, V. (2009). De ontknoping van de nodale oriëntatie: op zoek naar randvoorwaarden en kritische factoren. *Justiële verkenningen, 35* (1), 78-92.

Van Sluis, A., Marks, P, & Bekkers, V. (2011). Nodal Policing in the Netherlands: Strategic and Normative Considerations on an Evolving Practice. *Policing: A journal of Policy and Practice.* doi: 10.1093/police/par045.

Verhofstad, P.G. (2008). ANPR: toekomt voor een intelligente camera. *Het Tijdschrift voor de Politie, 70*(10), 4-8.

Welten, B.J.A.M. (2006). *De nodale oriëntatie en de discussie over privacy.*

ZHP (2009). *Korpsvisie 'Politiewerk in de haven': Tien punten aan de horizon.* Zeehavenpolitie Rotterdam.

SECURING THE LEGITIMACY OF SURVEILLANCE: AUTOMATIC NUMBER PLATE RECOGNITION IN DUTCH POLICING[1]

*Charlotte van Ooijen and Meike Bokhorst**

'We gave a presentation in the court of law, in which we showed a virtual oversight of a map. We used little light bulbs to show our observations of the historical data, ANPR and camera footage. That way we could, so to speak, let them drive towards the scene of the crime. Well, that made quite an impact, also in the court of law. That's the beauty of it. ... That's the power of combining your data. ... So far, I haven't had a single question about the acceptability of this data to the court of law. We simply explained in the official report how we got it, how it's made available and this is simply accepted. No questions. None at all. No.'

1 INTRODUCTION

The quote above by a Dutch police officer illustrates how the police can use Automatic Number Plate Recognition (ANPR) to retrospectively track the movements of a murder suspect. This surveillance technology seems to have found its way into the Dutch practice of criminal investigation and criminal justice. The apparent lack of questions about the acceptability of the mentioned application might suggest that either it is considered to be legitimate or that the police actors involved just don't care about legitimacy. In our view, both conclusions represent a much too limited view of legitimacy. Legitimacy-related questions about the application of ANPR could be: 'Does catching a handful of criminals justify the 24/7 scanning of all vehicles passing a particular ANPR-camera?'; 'Which information should be allowed to be connected to the registered number plates?' or 'Why should innocent citizens' data remain stored in police databases?' Legal theory and moral philosophy could be helpful to answer these kinds of questions. In this chapter, however, we take a different approach towards legitimacy

1. The authors would like to thank TMC Asser Press and Springer for granting permission to republish parts of Van Ooijen (2011).

* Charlotte van Ooijen (C.W.vanOoijen@tilburguniversity.edu) is a doctoral candidate at the Tilburg Institute for Law, Technology, and Society (TILT) of Tilburg University and is an expert member in Working Group 1 of the Living in Surveillance Societies COST Action. Meike Bokhorst is a doctoral candidate at the Tilburg School of Politics and Public Administration (TSPB) of Tilburg University and works as a research fellow at the Scientific Council for Government Policy (WRR).

issues. Instead of asking these kinds of normative questions ourselves, we are interested to learn whether these questions are posed by the actors involved in ANPR policy-making, and how these are addressed. In other words, how important is legitimacy in police practice and how do police actors deal with it? Consequently, the central research question in this chapter is:

To what extent do the police make an effort to secure the legitimacy of their ANPR policy and, what strategies do they deploy in doing so?

From both a moral and a practical point of view the issue of legitimacy is of relevance to the police. From a moral point of view, public organisations ought to strive for legitimacy. It would be amoral, at least in democratic societies, to exercise power without legitimately established rules, shared values, political acknowledgement and social acceptance. Legitimacy sets boundaries and obligations for the policies and actions of public authorities, thereby safeguarding citizens' fundamental rights. The issue of legitimacy is especially relevant in situations where the public organisation interacting with citizens is very powerful (Zouridis, 2007). The modern police force can be qualified as such for two reasons. First of all, their powers include the monopoly of the use of force that they share with the army. This right and ability to use force against citizens distinguishes the power of police from that of other public organisations like the Tax Authorities or the executive arm of the Dutch Ministry of Infrastructure and the Environment.[2] Secondly, a powerful organisation can become even more powerful once it has incorporated information technologies into its policies and practices (Zuurmond, 1994). This is clearly the case where the police are concerned, given the widespread adoption of nodal orientation, intelligence-led policing and technology-led policing (Bekkers & Van Sluis, 2009a; De Pauw, Ponsaers, Van der Vijver, Bruggeman & Deelman, 2011; Van Sluis, Marks & Bekkers, 2011). Besides their moral obligation to strive for legitimacy, the police have a practical drive to do so. They simply need legitimacy in order to survive. Vedder (2007a, p. 198) calls this pursuance of legitimacy 'for reasons of political efficiency and effectiveness'.

Securing legitimacy, however, is far from an easy task because it is an abstract concept (Zouridis, 2009, p. 293). Yet legitimacy *is* addressed in the practice of policing. This occurs even literally in the 'Police in Evolution' report,[3] in which the mission, vision and strategy for the Dutch police are discussed (Board of Chief Commissioners, 2005). This report shows 24 instances of the word 'legitimacy' in 18 pages. This indicates that the issue of legitimacy is of some practical relevance

2. This organisation is known as Rijkswaterstaat in the Netherlands (the Directorate General for Public Works and Water Management).
3. In Dutch: 'Politie in Ontwikkeling'

to the Dutch police, which isn't surprising when considering Beetham's argument that all power structures seek legitimation (Beetham, 1991, cited in Sparks, Bottoms & Hay, 1996, p. 85; Zouridis, 2009, p. 293). This assessment, however, doesn't help to form a thorough understanding of the meaning of legitimacy for technological-led policing. In order to do so we analyse legitimacy issues regarding the use of ANPR in a local police context.

Exploring legitimacy issues in a local police context

This chapter discusses one of the largest and most prominent police forces in the Netherlands as far as ANPR is concerned; it is referred to as 'the Nodalville police'.[4] In this case study, we discuss how the Nodalville ANPR policy has evolved in terms of legitimacy. Additionally, we consider how the police actors involved have attempted to secure this policy's legitimacy. In this study, policy-making is considered to be a non-linear process in which formal decision-making at the managerial level and work floor decisions at the implementation level influence each other. In reaction to the top-down approach of policy-making, this can be called a bottom-up approach in which policy implementation is viewed as part of the policy-making process rather than a separate process (Hill, 2009, pp. 202-204). Therefore, both formal decision-making and operational experiences with ANPR are included in the analysis. This case study is an endeavour to understand how police actors involved in ANPR policy-making perceive and handle this surveillance technology. As such, methodologically, this study can be characterised as an interpretive-qualitative policy analysis (Yanow & Schwartz-Shea, 2006; Yanow, 2007), in this case from the perspective of the surveillant.

Data generating for this case study took place between December 2008 and April 2009.[5] Data generating methods included participatory observation, interviews and document study. Three key informants played a central role, because contacts with them are more extensive than with others, and they regularly point to new sources of information in the course of the research. Participatory observation included shadowing[6] two of the key informants, who work as policy advisors in the department responsible for all policy development concerning information technologies. Field notes were made of overheard conversations and discussions. The third key informant leads a division responsible for all ANPR data management. All requests for ANPR data from within and outside the police

4. The real name of this police force is not disclosed at their request. The imaginary name 'Nodalville' points to the prominent attention for the nodal orientation strategy in this police force's ANPR policy.
5. For a more elaborate account of the position of the researcher and the research methods used in this case study see Van Ooijen (2011).
6. On the method of shadowing see Geuijen, 't Hart and Yesilkagit (2007).

force are dealt with by this 'ANPR division'. Therefore, this key informant is able to provide adequate information about the practice of ANPR. In addition, the ANPR data system and its application were observed during two police actions. Besides conversation-style interviews with the mentioned key informants at the Nodalville police force, semi-structured interviews are conducted with twelve informants. A total of seven interviews took place with informants representing the Nodalville police force. The other five informants are either national police representatives or employees of the local public prosecutor's office. The local police informants were mainly asked to describe their involvement with ANPR and elaborate on successes and problems they encountered. The other informants were asked to provide information about the policy context in which the local police force operated and comment on their experiences with the local police force. The interview protocol relied on open-ended questions, leaving room for additional questions tailored to the specific interview situation. The document study involved policy chapters, work instructions and internal (memos, intranet postings, e-mails) and external correspondence (newsletters, publications and presentations) to see whether and how legitimacy issues are addressed.

In order to analyse how the Nodalville police deal with legitimacy issues, the concept of legitimacy is first explored in literature. In our discussion of legitimacy, we distinguish legal, political, ethical and social dimensions. We structure ideas about the correspondence, whether desirable or not, between these four dimensions and their relative weight. After this discussion on the concept of legitimacy, we move on to the theme of nodal orientation. Our assessment is that this police strategy, which lies at the heart of innovations in the use of ANPR, raises some questions regarding legitimacy. In the subsequent presentation of the case study, first the functionality and applications of ANPR in the Nodalville police region are described. Then, we address the question of what the police actually do in order to be legitimate. To what extent and using which methods do the police pay attention to the four dimensions of legitimacy? Finally, the conclusions are presented together with a reflection on the importance of legitimacy in the practice of policing.

2 Legitimacy as a multidimensional concept[7]

Legitimacy is a key concept in the political sciences and public administration, but also an 'essentially contested' concept, which has sparked debate for almost as long as politics itself. Many use the term legitimacy and generally know what it means, but it remains a difficult concept to comprehend. What exactly is legiti-

7. This theoretical analysis of the concept of legitimacy is based on Meike Bokhorst's forthcoming PhD thesis.

macy? Which factors determine legitimacy? The lack of clear, univocal and unanimous answers to these questions results in some scholars narrowing down the concept, giving preference to related or underlying concepts (e.g. credibility, trust or support) or avoiding the term legitimacy altogether because it is complicated and 'it will never become clear anyway'. Others remain fascinated by attempts to make legitimacy more comprehensible. We prefer to make the concept more understandable by analysing different dimensions of legitimacy.

The concept of legitimacy was widely used from the nineteenth century onwards. The word itself appeared in the French political lexicon at the beginning of the nineteenth century and was then used only to describe a political system that is primarily determined by the law (Rosanvallon, 2008). Many dictionaries, like Webster or the Oxford English Dictionary, define legitimacy as 'legality' or 'conformity to the law'. But dictionaries give exactly the same definition for 'legality'. However, these two terms are not synonyms. Legality may be the literal translation of the term 'legitimus', but the development of legitimacy theory originated with the distinction between legitimacy and legality. By the thirteenth century, Thomas Aquinas had already distinguished illegal and illegitimate forms of tyranny. A tyrant can promulgate a law legally, but if that law is not approved by those to whom it applies, it will not be deemed legitimate. The idea of legitimacy stems from Roman law and is conveyed by the medieval argument of '*quod omnes tangit ab omnibus approbatur*': what touches all must be approved by all (Merquior, 1980, p. 3).[8] 'Recognition by all' thus forms an essential political dimension of legitimacy as well as a distinction with the legal term 'legality'. The law serves as an important instrument to codify what is recognised by all and to give it legal validity or strength. Legality thus is not a synonym, but one of the dimensions of or conditions for legitimacy.

There are many definitions of legitimacy in use, which have been heavily debated among political and social scientists (Barker, 1990; Beetham, 1991; Black, 2009). From these definitions and discussions, it can be concluded that legitimacy is a normative as well as a descriptive concept, one that has both an informal and a more formal (intersubjective or objectified) side. So we need a definition that combines those different aspects. Beetham's definition (Beetham, 1991) is a good example and is often and extensively used. According to Beetham, authority is legitimate when:

- it is acquired and exercised according to established rules (legality);
- the rules are justifiable according to socially accepted beliefs about (i) the rightful source of authority, and (ii) the proper ends and standards of government (normative justifiability);

8. Thinkers like William of Ockham, Grotius, Hobbes, Pufendorf, Locke and Montesquieu also contributed to the early theory development of legitimacy.

- positions of authority are confirmed by the express consent or affirmation of appropriate subordinates, and by recognition from other legitimate authorities (legitimation).

In this chapter, legitimacy is briefly defined. According to Beetham, it is the extent to which power is justified, legal, politically approved by all and accepted by subordinates. The interesting yet complex characteristic of the concept of legitimacy is its multidimensionality. One can state that the formal legal and political dimensions are each a necessary, but insufficient condition for legitimacy. Democratic constitutional states require laws for legal justification and politics ought to represent citizens' wishes. The necessity of informal ethical and social dimensions are matters which are more under discussion here. It is clear, however, that the lack of social and ethical legitimacy will soon affect the legal and political dimensions of legitimacy. Without justice, there can be no legality, and without social acceptance, there can be no political recognition. Indirectly and in the long run, the ethical and social dimensions seem to be necessary for legitimacy as well. It is debatable whether one can still speak of legitimacy if one of these four dimensions is missing. One can argue, however, that the higher an authority scores on each of the separate dimensions, the stronger its legitimacy is. This becomes especially clear when the four dimensions are presented in the following diagram:

Figure 1 Maximum legitimacy

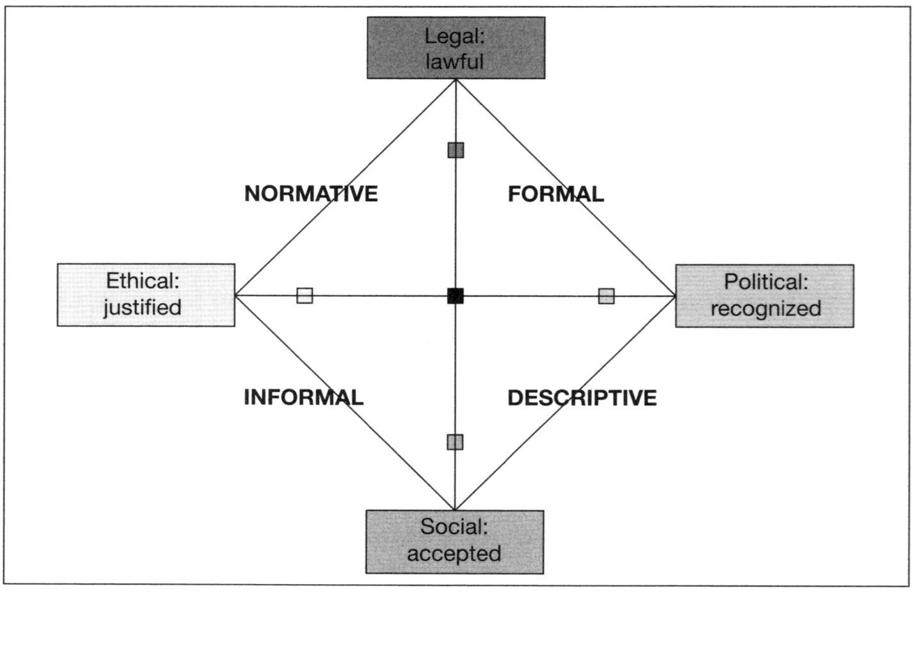

This diagram shows the level of legitimacy of a certain authority by determining the extent to which each dimension is met. By connecting the positions on the two axes a 'legitimacy diamond' appears, indicating the level of legitimacy. Figure 1 is an example of maximum legitimacy because the authority meets all four dimensions of legitimacy.

Relation between the dimensions of legitimacy

Acceptance, justice, legality and recognition by all citizens are four important dimensions of legitimacy. The question that remains is how they relate to one another and how they each contribute to legitimacy. A first possibility is a political system that respects human dignity, is recognised by all and is trusted by everyone. However, the majority of people will not consider the system to be legitimate if it does not function in accordance with the law. Every so often in a democracy, governments, ministers, members of parliament, mayors, aldermen and other political representatives resign after abusing their authority, unjustly using public means or exceeding important rules of integrity and trust. Ultimately, all citizens are equal before the law and administrators cannot create and enforce laws credibly and authoritatively if they do not comply themselves.

A second possibility is a political system that respects human dignity, acts in accordance with the law and is trusted by the people, but is not recognised by all and does not have a political mandate, will not be considered legitimate either. Leaders of unrecognised states or dictatorial regimes, for example, have difficulty gaining recognition, both within their own country and abroad. Even if the leaders have the power to occupy the seat of government and put their rivals in prison after a lost or manipulated election, they will still be illegitimate in the eyes of other states and without a clear political mandate.

A third possibility is a political system that respects human dignity, acts in accordance with the law, has once been recognised by all, but is not accepted by citizens. Such a system has a legitimacy problem that will only grow if the acceptance and trust decrease and remain fundamentally low. If the system does not succeed in solving this problem, it risks losing its political mandate in the next election because growing numbers of voters could be tempted to join protest parties and result in a decrease in voter turnout or participate in massive protests. For example, the political trust in Belgium has been so low lately that the tenability and rationale of the country's future is under discussion. It is difficult to measure and interpret issues like acceptance, trust and support. Since the degree of political trust fluctuates immensely and differs between groups, periodic elections are usually the most reliable and stable measure to determine political support and acceptance.

A fourth and final possibility is a political system that acts in accordance with the law, is recognised by all, and is trusted and obeyed by all, but does not respect principles of justice such as human dignity. Who would call such a system fully legitimate? Perhaps, for example, regimes that violate the rights of minorities or do not recognise human rights, like the freedom of press. Of the four dimensions of legitimacy, the principles of justice remain the most subjective, but they are also the most fundamental. Ever since the Second World War, international organisations have sought to politically objectify human dignity and other principles of justice, embed them legally and make them verifiable, as they did in the Universal Declaration of Human Rights (1948).

But how are these different dimensions of legitimacy related to the police strategy of nodal orientation? We will try to answer this question by describing the policy context of nodal orientation and analysing the ANPR policy of the police.

3 Policy context: nodal orientation

Nodal orientation as a new police strategy

In 2005, the Board of Chief Commissioners[9] of the Dutch police published a new vision memorandum on policing. The 'Police in Evolution' report portrayed a reorientation of the task and position of the police. The presented mission, vision and strategy were not supposed 'to do more – or less – than offer a course for developments to follow with a view to tailoring police tasks as closely as possible to society's requirements' (Board of Chief Commissioners, 2005, p. 12).

'Nodal orientation' was highlighted as a key concept in the new police strategy. Derived from the work of Manuel Castells' on the network society (Castells, 2000), the police distinguished a space of flows alongside a space of places, urging them to take up a new strategy that would concentrate on the nodes and flows constituting the space of flows. In their own words: 'The nodal orientation (''infrastructure policing'') leads to surveillance of the infrastructure, or rather, the flows of people, goods, money and information that use the infrastructure to move from one place to another' (Board of Chief Commissioners, 2005, p. 78). In order to realise the surveillance of flows, the police needed to develop information-driven working methods and secure the exchange of information with partners in the safety domain (Board of Chief Commissioners, 2005, p. 80). This emphasis on the importance of information encompassed a more technology-intensive way of working, which explicitly included the use of 'catch scan technology' where observations and registrations of people and vehicles are compared with a

9. Raad van Hoofdcommissarissen.

wide range of databases (e.g. unpaid fines, stolen vehicles, missing number plates, known suspects)' (Board of Chief Commissioners, 2005, p. 79). In subsequent policy documents this technology is predominantly called Automatic Number Plate Recognition (ANPR) and is as such referred to in this chapter.

Bekkers, Van Sluis and Siep (2006) emphasise that this particular interpretation by the Dutch Police of the concept of nodal orientation is rather unique, and should not be confused with nodal policing. In academic circles the latter concept usually points to the police as one of the nodes in networks of security oriented organisations (see Shearing, 2005). In a more recent discussion of nodal orientation in Dutch policing, however, Van Sluis, Marks and Bekkers (2011, p. 1) state that 'the Dutch interpretation of the nodal orientation combines a perspective on policing in infrastructural networks with one that considers the police as a player in (social) security networks'. Apparently, the meaning of the concept has evolved since its introduction in police circles in the 'Police in Evolution' report. Legitimacy problems arising from nodal orientation have been debated in both academic and police circles.

Legitimacy problems as a reason and a result

In a review of an article by Bekkers and Van Sluis (2009a), where they discuss the state of affairs regarding nodal orientation in Dutch policing, Bruggeman (2009) expresses three matters of concern which all point to legitimacy problems. First, he questions how far the police may and should go concerning the 'random' application of nodal orientation, especially in the light of privacy issues. Second, Bruggeman reacts to the assessment of Bekkers and Van Sluis that the current manner in which nodal orientation functions can be characterised as 'island innovation', a collection of as yet unconnected projects. As such, nodal orientation has so far not proven to be a new police strategy, but is merely useful for framing new practices and re-framing existing ones. Bruggeman (2009) wonders whether this use of the concept has any future at all in the light of ongoing technological developments. He argues that technological possibilities, such as the connection of Radio Frequency Identification (RFID) to the internet calls for more debate on technology-led policing in police circles. Consequently, he casts doubt on whether further investments in nodal orientation can be legitimated. The third issue raised is a need for self-regulation by the police to uphold the democratic quality of nodally oriented policing. In conclusion, Bruggeman states that once these three issues have been addressed, this will stimulate a more lasting acceptance among citizens. Summarising Bruggeman's concerns, he identifies legitimacy problems in terms of legality, moral justification and social acceptance. In subsequent responses, (Bekkers & Van Sluis, 2009b) as well as in a more recent article (Van Sluis, Marks & Bekkers, 2011), the authors affirm the importance of citizens' privacy as well as the issue of democratic accountability of nodally oriented policing. In addition, they emphasise that nodal orientation isn't merely a matter of technology, but has political relevance as well and as such it needs political-democratic legitimation (political legitimacy).

These are remarkable findings when reassessing the reasons why a new vision and an operational concept were needed in the first place. Some of the reasons which led the police to develop the strategy of nodal orientation can be seen as possible threats to legitimacy.

1. Decreasing effectiveness:

> The police signal a decreasing effectiveness in their task of safeguarding society and fighting crime, because of several changes in society, such as for example, individualization and increased mobility. The police themselves realise that 'efficiency and efficacy are now important pillars for the legitimacy of policing' (Board of Chief Commissioners, 2005, p. 25). Consequently, insecurity about what police measures can attain in contemporary society may infringe upon the right to use them in the first place.

2. Decreasing public support:

> 'There is a feeling that the police should perform better. Wherever the expectations and performance levels are difficult to reconcile, citizens become dissatisfied, they lose confidence in the police and the police lose their legitimacy. The police aim to continuously optimise their performance and, at the same time, their public image. Part of this process involves the profession forming a shared, coherent view on what will contribute to promoting safety. The present document contains the guiding principles of such a view, expressed as a mission, a vision and a strategy for the coming years.' Board of Chief Commissioners (2005, p. 21).

Apparently, the potential loss of legitimacy could be countered by developing a new strategy, according to the police. The respective contributions by Bekkers and Van Sluis (2009a; 2009b), and Bruggeman (2009), however, reveal that the strategy of nodal orientation seems to pose some legitimacy risks of its own.

This apparent tension of nodal orientation solving and causing legitimacy problems at the same time, can be understood when considering the scale of legitimacy. A distinction between two scales of legitimacy can be made: occurrent legitimacy on the one hand and dispositional legitimacy on the other. The latter refers to the legitimacy of an organisation in its entirety, whereas the first involves the legitimacy of a particular activity (Vedder, 2007b, p. 9). These two scales may be at odds with each other in two extreme ways. First, a legitimate organisation can perform non-legitimate actions. Secondly, there may be a non-legitimate organisation performing legitimate actions. Bearing in mind the multidimensional nature of legitimacy, it is unlikely for either of these scenarios is to be the case here. Regarding dispositional legitimacy, we will probably not find the Dutch Police to be entirely legitimate or unlawful. Indeed, the Dutch police organisation is facing some legitimacy problems, but this does not mean all legitimacy has disappeared. As far as occurrent legitimacy is concerned, it is also very questionable as well to assess whether nodally-oriented police actions are either completely legitimate or unlawful. Yes, nodal orientation poses a threat to citizens' privacy, but this doesn't imply that it is unlawful. It is more likely for one or more legitimacy dimensions of the police organisation to be at odds with one or more legitimacy dimensions of nodal orientation.

Nodal orientation, in any case, places the police in a difficult position. They may lose organisational legitimacy if they don't adopt this strategy, but run a risk of damaging occurrent legitimacy if they do. How do the police deal with this tension between legitimacy and non-legitimacy? How can they secure an equal balance of legitimacy if this is at all possible? In the next section, we discuss a case of nodal orientation in police practice: the use of ANPR by the Dutch Nodalville police force.

4 Case study: ANPR in police surveillance

Police surveillance in action: from hotlists to direct pursuit and further investigation

ANPR involves more than just cameras taking pictures of number plates. Since its first application in Dutch policing in the nineties by the National Police Services Agency[10] it has evolved into a complex system that can convert traffic data into valuable police information (Politie Rotterdam-Rijnmond, Korps Landelijke Politiediensten & VtS Politie Nederland, 2008). We emphasise 'can', because, as will be demonstrated, not all data actually amounts to information.

Technologically speaking, the Nodalville ANPR system consists of cameras for data gathering, databases for data storage, a network connection for data access and software for data linkage and analysis. We briefly highlight some basic aspects of each of these technological components. The Nodalville police force owns both fixed and mobile cameras. The fixed cameras cover the major entry and exit roads and gather data for 24 hours a day, 7 days a week. The mobile cameras in police vehicles are switched on more irregularly. What data do these fixed and mobile cameras exactly gather? A camera automatically takes pictures of the number plates of all passing vehicles. The underlying software then converts these digital photos into alphanumerical data. Both the picture files and

10. This first application in policing was predominantly aimed at traffic checks.

the alphanumerical number plate data are stored along with data on location and time. The network connection then enables an authorised police-officer at the office to access this data within seconds. From one fixed location alone on the average over 80,000 passages are registered per day. At the time this case study was conducted, all data was stored for four months and permanently deleted afterwards, except for the data which in the meantime had been selected for further police investigation. A software program helps the police to enrich the millions of stored ANPR data records in order to obtain information that could be of value in different kinds of police processes. The program creates hotlists of number plates with which 'something's up', to use the words of a police officer. At this point in time, the police define the criteria for inclusion on a hotlist by looking at the car owner's past or present offences. Examples range from unpaid speeding fines, alcohol related offences and drug trafficking to wanted felons. By linking police data on offences to data on car owners, hotlists of number plates warranting attention are generated.

Basically, the hotlists can be used in two types of applications, either direct pursuit or for further investigation. Direct pursuit involves comparing number plates to a previously composed hotlist in real-time to take immediate action. The Nodalville police organise special action nights to actually remove the 'hits' from the anonymous traffic flow. On these occasions, police motorcyclists are present at one of the fixed camera locations. Upon a signal from their colleague who is watching the ANPR system, they drive up to a particular car and escort it to the side of the road where police officers are awaiting to deal with the particular offence. This, for example, involves collecting fines, checking alcohol abuse or making an immediate arrest. Until 2007 the Nodalville use of ANPR was limited to these types of applications, which are aimed at traffic control. This focus is reflected in the police organisation where ANPR is handled by the traffic police department. This changed in the course of 2007 when the responsibility for ANPR policy and organisation shifted to tactical-operational and strategic information divisions within the police force. These divisions take care of passing vehicles that are not registered on any hotlist, the so-called no-hits. They do remain saved in the ANPR system, because of their possible relevance for the second type of application: further investigation. This means that a picture, licence plate data, location and time are registered for all vehicles passing ANPR cameras in the Nodalville police district.

The reasoning of the Nodalville police is that all ANPR data, both 'hits' and 'non-hits', may prove to be of use later on for crime fighting purposes. Specific queries help the police to allow the necessary information emerge from the huge amount of stored data. As we mentioned in the introduction of this chapter, the police have used the ANPR system, for example, to retrospectively track the movements of a murder suspect. Another way of using ANPR data for further investigation is the analysis of movement patterns of a particular group of people. The Nodalville police have, for example, compiled a hotlist of house burglars. On this list there are names and number plates of people who are known to have been convicted for, suspected of or otherwise involved in burglaries. This list is updated weekly and used to find suspects of particular burglaries, and to enable the police to catch burglars red-handed. Finding suspects is done by geographically plotting committed burglaries and hotlist movements. Also, the time of the burglaries and the time at which the fixed camera locations were passed are taken into account when determining which cars could be connected to which burglaries. At the same time, the system allows for exclusion of particular suspects. Furthermore, movement patterns are analysed to determine at what time a burglar returning home is likely to pass by one of the camera locations. This allows the police to position a field officer at the scene so they can single this person out of the traffic stream, possibly finding the stolen goods in the trunk of the car.

The Nodalville police ANPR policy

The application of ANPR for the purpose of further investigation confronts the Nodalville police with questions concerning the boundaries of the use of this technology. Framed more positively, since the expansion of ANPR equipment in 2007, the police have speculated about what else they can and may use ANPR for, as several policy documents reveal.[11] The discussion about the limits to ANPR has moved on since 2007, because the Nodalville police have expanded their range of ANPR devices and software and looked into applications for further investigation that until that time weren't customary. The police initially focus their attention on the legal background, but later reports show contemplations about social support and political reactions. Between 2007 and 2009 the Nodalville police demonstrated an ambiguous approach towards the legal dimension of legitimacy. In several policy documents the Nodalville police assess that there is no law in the Netherlands which clearly determines how the police may or may not use ANPR. Nodalville policymakers actively investigated which laws are applicable and came to the conclusion that the current legal framework doesn't provide sufficient guidance on what course of action should be followed. This assessment, however, doesn't pose a reason why not to develop any course of action at all. In the lack of clear regulation, the legal department develops their own privacy policy in order to provisionally give direction to their ANPR actions. At the same time, the analysis of policy documents reveals that legality is-

11. These involve four internal policy documents which were accessed confidentially for research purposes. Citations from these documents are therefore referred to by mentioning 'Nodalville police policy document' instead of their exact source. These documents were written between 2007 and 2009.

sues are considered to be a risk as well. The police express their concern that the development of legal norms and political discussion may put a stop to the originally designed and enacted plans. This is especially the case for possible violations of citizens' rights to privacy.

In order to overcome this legal insecurity, the police have developed an instrumental concept of legality. They don't wait for legal security to arrive by itself, but actively engage in discussions at the Ministries of Interior Affairs and Justice in order to help create legal possibilities. As one informant formulates it: 'We want to prevent having a suspect in front of a judge who then considers the ANPR data to be illegitimate evidence'. The Nodalville police actively engage in other discussion groups as well. They, for example, have representatives in all nationally organised police work groups on ANPR. Still, this doesn't automatically mean that the political dimension of legitimacy is honoured in the strict sense of the word. Informants say that their reasons for participating in these groups are mainly to educate other police forces on how they should handle ANPR and to influence the discussion on legal norms:

> 'We now discuss these matters in order to prevent being stopped later on.'[12]

> 'The connection to the national program agency has few consequences for the further development of the regional project. The regional project will supply the national program with information on how cases are solved in Nodalville'[13]

Social support for ANPR did not receive the Nodalville police's attention until 2008. Policymakers actively follow the ongoing political and societal discussion on ANPR. For this purpose, they consult research reports, follow the news media and consult online forums where citizens express their opinions on the matter. Policymakers state they take what some actors have to say more seriously than others. Nonetheless, they also listen to actors they find to be 'less relevant', but more for the sake of listening than for the actual outcome. The Dutch Data Protection Authority (*College Bescherming Persoonsgegevens* (CBP)), for example, did receive a formal reply from the Nodalville police on one of their reports, but it was granted little importance in the actual policy-making process. According to the Nodalville police policymakers, this strategic consideration, however, could strengthen the police's consideration of social developments. In addition, the Nodalville police aim to limit society's awareness of ANPR through their conservative communications policy. There have been press releases about the direct pursuit application, but these don't give away details about the technology.[14] They

12. Quote from interview with Nodalville policymaker.
13. Nodalville policy document is a fictitious name given to the document by the researchers.
14. According to press releases found on the Nodalville police's intranet.

merely mention the results of a particular action, such as how many fines were collected at a certain location. On certain occasions some information about crime fighting applications is provided, but only in very general terms and on the request of journalists.

For most informants it is quite clear that ANPR already increases the effectiveness and efficiency of the police force and will do so even more in the future through expanding applications. Informants and documents usually refer to success stories of the direct pursuit application in law enforcement to support their claim, and use this argument to legitimise further applications of ANPR. The legitimacy of a particular police action (the direct pursuit application) is used to refer to the legitimacy of the police organisation as a whole. 'Policing becomes more effective', was a phrase expressed more than once. In addition, this claimed dispositional legitimacy is used to justify yet another police action – the further investigation application. As a result, the provided justification is not a sound one. In addition, the Nodalville police increasingly emphasise the value of proactive policing as a supplement to reactive policing. According to a policymaker, proactive policing may help the police because this approach could diminish reactive policing: 'Now only about two out of ten cases can be handled'. One of the values of proactive policing, therefore, could be to gain in efficiency, according to the police. The concept of nodal orientation is also often used as a means of justification: 'ANPR is efficient, it's effective, and of course it's nodal orientation.' Most informants state this without properly explaining what nodal orientation entails and why it is important. The power of the concept of nodal orientation is that it offers a strategic perspective by reframing existing practices and projects and framing new practices (Bekkers & Van Sluis, 2009a). This is precisely what has happened in the police region of Nodalville. They started using ANPR shortly before the 'Police in Evolution' report appeared and ideas about nodal orientation could be passed on to the Dutch police forces. This police force has been using ANPR for law enforcement purposes since 2004 but is gradually expanding the range of applications to innovative ways of crime fighting. At the same time, the number of ANPR cameras owned by the Nodalville police has substantially increased, resulting in more data gathering. Using ANPR at infrastructural nodes doesn't automatically imply a strategic repositioning of the police. Van Bruul, Damkat, Van Dijk, Frenken and Van Haasteren (2008) argue that in most cases this is no more than a technical approach to traditional police work (Bekkers & Van Sluis, 2009a, p. 52). Even if this were the case in Nodalville, ANPR still is a very powerful tool.

Essentially, the above-mentioned explorations made by the Nodalville police force have to do with developing a legitimate ANPR policy. A provisional counterpoint in this policy process has been the blame expressed by the CBP in January 2010. The CBP concludes that two Dutch police forces violated the law knowingly and willingly by saving data about no-hits (College Bescherming Persoonsgegevens, 2010). The subsequent order from the CBP to erase no-hit data from all police records also applies to the Nodalville police.

Case analysis from a legitimacy point of view

In this case study two important policy changes are observed that are connected to a changed perception of legitimacy by the Nodalville police. The first change involves the transition from state regulation to discretionary drive in 2007. During the second change, at the start of 2010, discretionary drive was replaced by socially restrained ambition.

The first change was initiated by the broadening of ANPR applications from traffic control to innovative crime fighting. The interaction between the Nodalville field officers and strategic policymakers instigated this expansion. The field officers used the extra equipment, which is at their disposal between 2004 and 2007, not only to conduct more traffic controls, but to take initiatives to analyse the collected data for crime fighting purposes as well. At the same time, strategic policymakers work on the development of the nodal orientation strategy featuring a big role for ANPR. The interaction between both developments allows the Nodalville police to further develop ANPR policy and practice with discretionary drive. This development style could be characterised by Nodalville police's drive to explore and implement the possibilities for the execution of police tasks that are generated by technological developments. Meanwhile, an ambiguous discretion is at play. On the one hand, the police force shows discretion in the sense that they interpret legal, political, ethical and social preconditions rather freely. The Nodalville police, for example, interpret the Police Data Act[15] in such a way that it covers the storage and processing of no-hits. On the other hand, the police force shows discretion by being very cautious in their external communication about the most innovative ANPR applications. As a part of the discretionary drive during this second phase (2007-2009), the Nodalville police started to pay more attention to legitimacy because it had become more debatable for the two reasons we mentioned in the section about nodal orientation: decreasing effectiveness and decreasing public support.

In summary, the expansion in ANPR applications in 2007 causes lack of clarity in the area of legitimacy. It is very uncertain to what extent and which new applications are to be legitimised. Also, some questions remain in the ethical area, because the Nodalville police are insufficiently able to demonstrate whether the innovative application of ANPR helps them to fulfil the moral duty of protecting a changed society. Politically and socially speaking there is little or no attention for

15. 'Wet Politiegegevens' in Dutch.

the changed ANPR policy in Nodalville. At the same time, the legitimacy of ANPR applications in police traffic controls has been secured as a result of an increased effectiveness and efficiency within legal limits. Moreover, little social resistance is found towards these ANPR checks. The legitimacy diamond below illustrates this.

Figure 2 Discretionary drive in Nodalville ANPR-policy

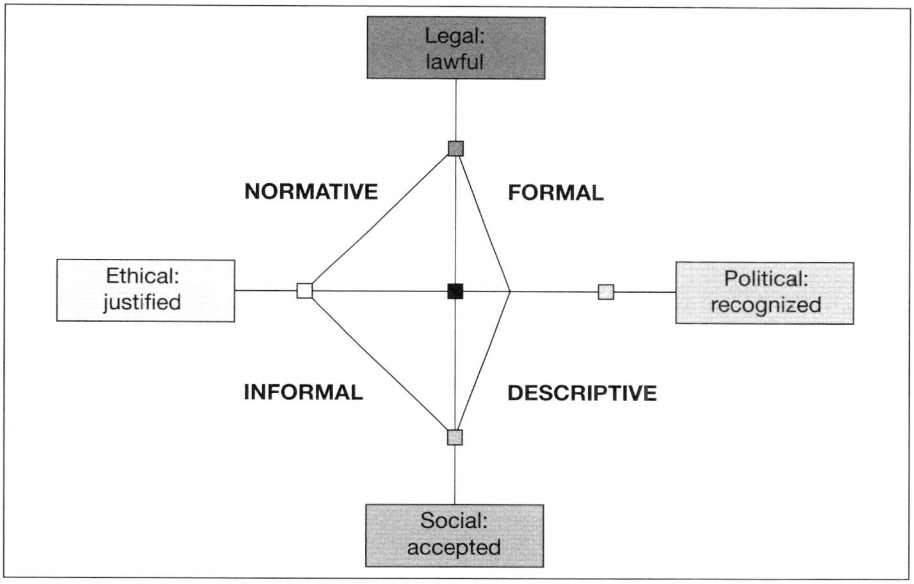

More clarity about what is legitimate concerning ANPR applications, and especially about what is not, arised at the start of 2010 after the publication of the CBP report. The CBP's report reprimands the Dutch police force, warning that it is neither socially nor legally acceptable to record no-hits. Consequently, the Nodalville police were forced to deviate from their discretionary drive to restrained ambition. The investments in ANPR equipment, organisational support and contact with other police forces, the national police organisation and the Dutch Ministries of Interior Affairs and Justice actions have paid off because there is increased political support for further development of ANPR. The coalition agreement of the Dutch cabinet of the day states that: 'Vehicle recognition will be used to facilitate the prevention, investigation and prosecution of crimes and the enforcement of fiscal obligations. To this end, wide use will be made of automatic number plate recognition systems' (Dutch Government, 2010, p. 56). In 2011, the Minister of Security and Justice made this plan more concrete by proposing a new law to parliament 'to allow the police to log information, obtained by ANPR for a period of four weeks, for criminal investigation purposes after a criminal act has been committed or after criminal charges have been pressed'

(Van Sluis, Marks & Bekkers, 2011, p. 7). Important from the point of view of moral justification is that the new law will only be prolonged after its trial period of three years if the use of ANPR data proves to be effective for the criminal investigation process. In short, even though political support has increased significantly, the ANPR policy still lacks legal, ethical and social legitimacy, which is illustrated in the legitimacy diamond below.

Figure 3 Restrained ambition in Nodalville ANPR policy

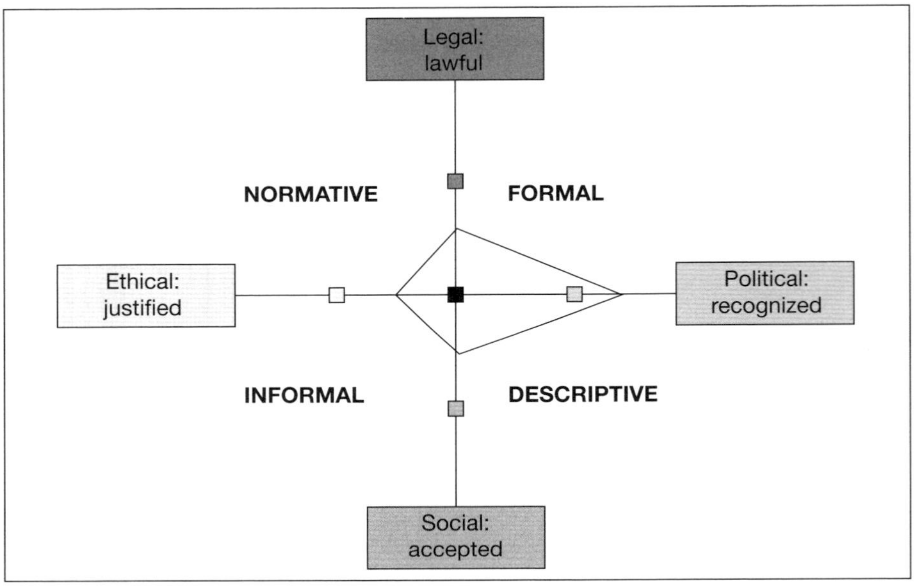

5 Conclusion

To sum up the arguments and findings, we will reflect on the essential points and aspects. The initial research question was: to what extent are the police making an effort to secure the legitimacy of their ANPR policy and what strategies do they deploy in doing so? The case of the Nodalville police shows that police policymakers are very sensitive about legitimacy issues, but mainly for pragmatic reasons. The legitimacy diamond we present as an analytical tool has helped to make clear which shifts take place in the legitimacy dimensions in the course of the years, and how these influence policy-making in the case of the Nodalville police force. In this case, the legal dimension of legitimacy plays an important role. The police's insecurity about the lawfulness of innovative ANPR applications has led to a legitimacy offensive towards other Dutch police forces, politics and society. Among other police forces and in a political working group, the police instigated moral awareness of not disregarding the ANPR potential to combat new crime (ethical legitimacy). At the same time the police emphasised, in the media, the effectiveness and efficiency to be gained from ANPR for policing while disregarding the privacy implications of innovative applications, and the extent to which people value their privacy in comparison to safety (social legitimacy). In order for the police to come closer to securing the legitimacy of this type of surveillance, they need to find substantial support for the moral claim of increased effectiveness and efficiency. The fact that the Minister of Security and Justice has ordered an evaluation of the effectiveness of ANPR for criminal investigation purposes, along with a new law for ANPR, is a good start to solving this legitimacy deficit.

This case is a typical example of the idea in the sociology of law that there is a gap between the law of the books and the law in action. Especially in case of fast technological developments it is difficult to develop a new legal framework in advance or simultaneously. First there is an (implicit) normative consensus within the police that the way they use the ANPR applications is justified largely due to instrumental gains. And, furthermore, the police are trying to convince politicians and the public that the legal framework has to allow for adaptations to the new technological possibilities. It could be that the police will discover that there is a lack of support in public opinion once it is too late. A way to bridge or to narrow the gap between the law of the books and the law in action is to explore and raise the normative consensus at an early stage, for instance, by organising a dialogue about guidelines for the use of ANPR with stakeholders outside the police, such as the interest groups for car drivers, protectionists of public safety, privacy activists, etc.

Besides legitimizing actions in the policy practice of ANPR, some work has to be done at the strategic level of nodal orientation. Some confusion in discussions about the legitimacy of nodally-oriented policing has arisen, because arguments referring to occurrent and those referring to dispositional legitimacy get intertwined. This is not very surprising given the ambiguous meaning of nodal orientation in Dutch policing, pointing to both surveillance of infrastructures (occurrent) and the position of the police organisation in security networks (dispositional). Further development of nodal orientation as a police strategy is necessary to gain more clarity about the legitimacy of its still existing practices, such as ANPR. For this purpose, both the police and academics should work together to make the relation between the occurrent and dispositional nodal orientation more explicit. We hope our findings will help to foster an understanding of the relation between both legitimacy scales of nodally-oriented policing.

References

Barker, R. (1990). *Political legitimacy and the state.* Oxford: Oxford University Press.

Beetham, D. (1991). *The legitimation of power.* London: Macmillan.

Bekkers, V., A. van Sluis & P. Siep (2006). *'De Nodale Oriëntatie van de Nederlandse Politie: over criminaliteitsbestrijding in de netwerksamenleving. Bouwstenen voor een beleidstheorie'.* Rotterdam: Center for Public Innovation, Erasmus University Rotterdam.

Bekkers, V. & A. van Sluis (2009a). Nodale politie in Nederland. Een tussenstand van zaken. *Panopticon. 30*(2), 49-54.

Bekkers, V. & A. van Sluis (2009b). Reactie op de nabeschouwing van Prof. dr. Willy Bruggeman. *Panopticon. 30*(2), 55-56.

Black, J. (2009). *'Legitimacy and the competition for regulatory share'.* LSE Working Papers 14/2009.

Board of Chief Commissioners (2005). *'Police in Evolution. Vision on Policing'.* The Hague: Board of Chief Commissioners / NPI.

Bruggeman, W. (2009). Nabeschouwing bij de tekst 'Nodale oriëntatie'. *Panopticon. 30*(2), 54-55.

Bruul, I. van, P. Damkat, W. van Dijk, W. Frenken & P. van Haasteren (2008). *'Leiderschap als knooppunt. Een studie naar de rol van leiderschap in de diffusie van nodale oriëntatie binnen de Nederlandse politie'.* Warnsveld: School voor politieleiderschap.

Castells, M. (2000). *The rise of the network society.* Oxford: Blackwell.

College Bescherming Persoonsgegevens (2010). *Politiekorpsen handelen in strijd met de wet bij toepassing ANPR.* http://www.cbpweb.nl/Pages/pb_20100128_anpr.aspx (17 december 2010).

De Pauw, E., P. Ponsaers, K. van der Vijver, W. Bruggeman & P. Deelman (Eds.) (2011). *Technological Led Policing.* Antwerpen-Apeldoorn-Portland: Maklu.

Dutch Government (2010). *Freedom and responsibility. Coalition Agreement VVD-CDA.* http://www.kabinetsformatie2010.nl/dsc?c=getobjects=objobjectid=127511 (15 April 2011).

Geuijen, K., P. 't Hart & K.Yesilkagit, Kutsal (2007). 'Dutch Eurocrats at Work: Getting Things Done in Europe'. In R.A.W. Rhodes, P. 't Hart & M. Noordegraaf (Eds.). *Observing government elites: up close and personal*: 131-159. Basingstoke: Palgrave Macmillan.

Hill, M. (2009). *The Public Policy Process.* Harlow: Pearson Longman.

Merquior, J.G. (1980). *Rousseau and Weber. Two studies in the theory of legitimacy.* London: Routledge & Kegan Paul.

Politie Rotterdam-Rijnmond, Korps Landelijke Politiediensten and VtS Politie Nederland, onderdeel NPI (2008). *'Automatic Number Plate Recognition, Naar een landelijke toepassing. Visiedocument ANPR'.*

Rosanvallon, P. (2008). *Counter-democracy. Politics in an age of distrust.* Translation of 'La contre-démocratie'. Cambridge: Cambridge University Press.

Shearing, C.D. (2005). Nodal security. *Police Quarterly* 8 (1), 57-63.

Sparks, R.F., A. Bottoms & W. Hay (1996). *Prisons and the problem of order.* Oxford: Clarendon Press.

Van Ooijen, C.W. (2011). 'Legitimacy Issues Regarding Citizen Surveillance: The Case of ANPR Technology in Dutch Policing'. In S. van der Hof, & M.M. Groothuis (Eds.), *Innovating Government*: 197-216. The Hague: T.M.C. Asser Press.

Van Sluis, A., P. Marks & V. Bekkers (2011). Nodal Policing in the Netherlands: Strategic and Normative Considerations on an Evolving Practice. *Policing* 5 (3), 1-7.

Vedder, A. (2007a). 'Towards a defensible conceptualization of the legitimacy of NGOs'. In A. Vedder (Ed.), *NGO involvement in international governance and policy: sources of legitimacy*: 197-212. Leiden: Nijhoff.

Vedder, A. (2007b). 'Questioning the legitimacy of non-governmental organizations'. In A. Vedder, (Ed.), *NGO involvement in international governance and policy: sources of legitimacy*: 1-20. Leiden: Nijhoff.

Yanow, D. & P. Schwartz-Shea (Eds.) (2006). *Interpretation and method: empirical research methods and the interpretive turn.* Armonk, NY: M.E. Sharpe.

Yanow, D. (2007). Interpretation in Policy Analysis: On Methods and Practice. *Critical Policy Analysis*. 1 (1), 109-121.

Zouridis, S. (2007). 'De legitimiteit van recht en bestuur: signalen voor legitimiteitserosie geanalyseerd'. In Cornelissen, E.M.H., P.H.A. Frissen, S. Kensen and T. Brandsen (Eds.) *Betoverend bestuur: legitimiteit, vitaliteit, meervoudigheid*: 93-116. Den Haag: Lemma.

Zouridis, S. (2009). *De dynamiek van bestuur en recht: over de rechtsstaat als bestuurswetenschappelijk fenomeen.* Den Haag: Lemma.

Zuurmond, A. (1994). *De infocratie: een theoretische en empirische heroriëntatie op Weber's ideaaltype in het informatietijdperk.* Den Haag: Phaedrus.

The political geography of public space. On criminalisation and punishment, privatisation, dispersion and exclusion

*Antoinette Verhage and Paul Ponsaers**

1 Introduction

The concept of 'community' has become more diverse and more complex in recent decades. This also implies that policing these communities has become more complex and that surveillance needs to adapt to new and broader sets of norms and rules. As a result of this diversification, the policing of public spaces has dispersed into several types of social control, carried out by a diversity of actors. These types of social control are summarised below, under four headings: criminalisation and punishment, privatisation, dispersion and exclusion. Each and every one of these types of social control has a specific objective, related to the method applied. In this contribution, we want to discuss these different methods of social control and make clear that they have resulted in a muddle of policing actors and approaches that have permeated public spheres.

Although policy strategies implicitly aim for an 'informalisation' of policing (through the use of more approachable or accessible styles of policing), it seems as if this diversity in social control varieties, on the contrary, results in an increased formalisation of policing and surveillance, or, as Jones and Newburn have called it, a long-term process of formalisation of social control (Jones & Newburn, 2002). In this contribution, we aim to analyse how this impacts on Belgium and reveal the questions that underlie these developments. The contribution will start by discussing the changes in social control and disruptions of social order, and elaborate on the adjustments of policing that result from it. This implies that we will look at the effects of these developments on the surveillant and the surveilled (the different groups 'inhabiting' public space), but also at the effects on policing of public order and public space as such. Essential in our view

* Antoinette Verhage works as Assistant Professor at Ghent University, Faculty of Law, Department of Penal Law and Criminology. She is co-director of the Research Unit Social Analysis of Security (SVA).
Paul Ponsaers is full time senior professor at Ghent University, Faculty of Law, Department of Penal Law and Criminology.

on the contemporary interpretation of policing public space is the observation that strategies are no longer exclusively police-oriented by nature.

2 Public space and social control

Public space implies areas which members of the population at large are free to enter or leave at their own discretion. The term 'public space' can be interpreted in many ways. In literature, sometimes the distinction is made between 'representational space' (appropriated space, space that is used by specific groups) and 'representations of space' (planned, ordered space, such as a courtyard or a park) (Mitchell, 1995). Public space belongs to citizens, can be accessed by everyone without restriction, and the interaction which takes place there is among persons of equal standing who generally do not know one another. For this reason, public space – the street, open-air markets, parks, squares and public facilities such as train stations and post offices – is the pre-eminent place where people interact with other people whom they do not know. Public space has the potential for an endless number of experiences, and as such presents many more opportunities than threats or insecurity (Van den Herrewegen, 2008). There are many pleasant activities in which to partake and the potential for communication is enormous, which means that public space is a sort of theatre, a stage, for social interactivity. And finally it is the place for active citizenship, where it is possible to voice protest and be heard by others. In other words, as Fraser states 'public spaces are absolutely essential to the functioning of democratic politics' (Fraser, 1990, quoted in Mitchell, 1995, p. 115). In short, the street has a liberating influence and is essentially characterised by diversity brought about by changing minorities and majorities. These different interpretations and notions of public space will be illustrated throughout this contribution.

Of course, the freedom enjoyed by any individual person in public spaces must be limited so as to respect the freedom of all other citizens. After all, it is important to always remember that others must (if they wish) be able to make use of that space, which by definition immediately implies the need for an open mind and flexible attitude as well as a number of basic rules. In this connection, Lyn Lofland (2006) set out a few principles for good public behaviour, such as a) co-operative mobility, so that everyone is able move around freely, b) polite inconspicuousness for a certain level of discretion and politeness, c) a desire to be an active spectator and listener by showing interest, attention and alertness, d) demonstrating restrained helpfulness through concern for others and, e) tolerance as regards diversity, i.e. not having any prejudices. As such there are indeed rules of etiquette for social interaction which can and must be applied in a respectful manner because the use of public space implies active participation in regulated social interaction.

It is in this context that informal forms of social control become important. In a long-term analysis, Spierenburg (2007) indicates that informal forms of social control take place at the neighbourhood level. He observes that during the sixteenth and seventeenth centuries informal and semi-formal forms of social control were still very much present and generally materialised through entertainment, gossip and mockery. However, he points out that these forms of informal social control hardly ever involved criminality, but rather had to do with regulating social interaction, i.e. abiding by the rules of etiquette in public spaces. People were accustomed to the fact that their neighbours were actively involved in everyone else's comings and goings. Again according to Spierenburg, the pride associated with the reputation of all respectable villagers was considered a collective asset which had to be protected.

It is generally presumed that internal social control within local communities decreased in line with industrialisation. However, based on various historical studies, Spierenburg argues that this process only began in a later period and that gossip and mockery continued to be how villagers reprimanded one another for quite some time. According to him, it is only since the Second World War that intensive forms of local social control definitively disappeared, that support for conventions encompassing more than just the fight against crime collapsed. For example, in this connection who still remembers the signs from long ago in public parks stating 'This public garden is under the protection of members of the public'?

Specifically, it is this second point that drew increasing attention from the authorities and became progressively more formalised after WWII. In fact, it evolved to become the pre-eminent remit of local police with the advent of the theme of so-called 'widespread petty crime' during the 1980s. When Hein Roethof submitted the interim report of the *Commissie Kleine Criminaliteit* (Commission on Petty Crime) in 1984, the prevailing policy theory in the Netherlands (which was also solidly taking root in Belgium) was that various forms of informal social control had disappeared (for example, as exercised by doormen in large collective housing units, tram and bus drivers, etc.) and that this void had to be filled by a good partnership between the police and the population at large so that the growth in petty crime could be reversed (Commission for Petty Crime, 1986). Aside from the question of whether this position was correct, it remained unknown what should be used to compensate for this diminishing effect of social control of those broad conventions encompassing more than purely the fight against crime. After all, it had been a long time since any behaviour transgressing these conventions and exhibited in public could be considered as formally punishable and subject to police intervention.

To a large extent this question has a lot in common with the problem described much earlier by Ferdinand Tönnies in his work *Gemeinschaft und Gesellschaft* (Tönnies, 1935). According to him, social interaction in a '*Gemeinschaft*' (community) is governed by shared moral codes covering appropriate behaviour by members towards one another and towards the community as a whole. Again according to Tönnies, there is seldom a need for external social control in this type of structure. However, in a '*Gesellschaft*' (society), there is no longer the same level of shared moral values, individuals act out of self-interest, and there is a high risk of conflict, which means that the need for external social control increases. This line of thought was echoed by Durkheim in his *The Division of Labour in Society* (Durkheim, 1947) when he stated that '*Gesellschaft*' should be seen as 'a progressive development of individualism, the dispersive effects of which can only be prevented for a time, and by artificial means by the action of the state, and that it is essentially a mechanical aggregate' (Durkheim, 1947, p. 146). Therefore, both of these classical thinkers believe in the assumption that local societies evolve from a more closed and homogeneous form to an open and more heterogeneous form. Conflicts are inherently tied to these more complex societies and they cannot be permanently solved by the actions of the state, in this case through the police.

Formal social control in the public space was traditionally interpreted as public order maintenance; this raises associations with one of the most classical police functions. Manifestations of this classic police task could be seen during the riots in London and Paris. What is happening today, is that – in addition to this classic police task – the public space is impregnated by all types of regulatory mechanisms that transcend the classic interpretation of policing. How can all these types of mechanisms be managed and who is supervising them? What does this imply for control and is there someone to guard the helicopter view?

The question we wish to answer in our contribution is as follows: precisely what political geography of public space is being applied to our late modern society or, to put it more simply, what methods of approach (or strategies) are being developed in our local, late modern societies to ensure the maintenance of the abovementioned broad conventions in public spaces? How is behaviour in public spaces channelled by the diversity of actors? Of course, the question here also concerns the role of the police in this context.

3 Welcome to the public space

If we look at the prevailing contemporary conceptualisation of the matter, the key question concerning public space essentially seems to boil down to the following: Are young people welcome in city centres? (Crawford, 2006).

It is true that to a large extent that the issue in question here focuses on city centres. These places are viewed as the pre-eminent locations where problems involving public spaces materialise and are concentrated (Goeminne, Enhus & Ponsaers, 2003). Looked from this angle, a city's centre is a very ambiguous place. Due to the mono-functional nature of our city centres, which almost entirely focus on consumption (Maciocco & Serreli, 2009), on the one hand they are an invitation, a sort of display case, for the majority of people who are specifically drawn to this consumptive behaviour. It is these people who want to go to the city centre and who are also welcome there.

Moreover, this mono-functional character as regards consumption is to a large extent further emphasised by the fact that the residential character of city centres is largely on the wane as a consequence of the urban flight seen over the past few decades (Ellen & O'Regan, 2010). People, in their capacity as users of the city centre, come and go in accordance with the rhythm of their consumption patterns, i.e. based on the opening and closing times in shopping districts. This therefore implies that the public space in the city centre is not the consumers' destination in and of itself and that they primarily use public space as a means of access to reach their consumption destinations (shopping malls, cinemas, entertainment districts, etc.). Mobility and accessibility have become the key issues in the management of our city centres, and this is clearly illustrated by parking garages under main squares, centralised terminals of bus and tram lines, as well as centralised train and underground train stations.

On the other hand, in this context we also study young people and their behaviour (Fielding & Innes, 2006). According to most people, it is them who have failed to adequately adopt broad social conventions and often violate them. For this reason they are considered responsible for hampering the regulation of public space, particularly in city centres. Young people are seen as creators of disorder who partially privatise public space because they physically occupy part of the public space and so put the boundary between public and private areas under pressure (Pleysier & Deklerck, 2006). However, it must also be pointed out that public space is a legitimate and important place for young people to engage in social activities (Malone, 2002).

Essentially, young people are viewed as 'bad' consumers who normally do not have enough financial resources to be considered (financially) 'complete' and 'worthwhile'. It is possibly not purely coincidental that in certain cases shopping malls in Britain now only grant access to young people if they can prove that they have money with them, or that at the end of the 1990s in Australia they were simply denied access to certain shopping centres, a practice which incidentally was labelled by the United Nations as a 'discriminatory policy' (Malone, 2002).

For many young people, public space is not only a means of access to reach their destinations, but it is also the destination itself, a place where they can meet. However, young people who use public space in that manner risk being labelled 'loitering teens'. For them, the city centre is the ultimate meeting place, a key location that is not only a nice place to be but also where they feel safe precisely because of the abundant public space. As regards members of the public, however, the mere presence of young people at a 'hangout' (for example, a park, as in research by Pleysier and Deklerck in 2006) is enough to generate a feeling of irritation or aversion.

In his conceptualisation of 'Reassurance Policing', Martin Innes (2005) strongly emphasises the notion of 'risk signals'. His analysis is based on the subjective issue of feelings of insecurity and for this he departs from the finding that feelings of insecurity should in many cases actually be understood as a fear of disorder. Here he bases himself on the results of the British Crime Survey (Wood, 2004) which indicated that three-quarters of respondents were of the opinion that anti-social behaviour was a 'very big' problem and that many believed that this sort of behaviour was significantly increasing. Young people were generally identified as the cause of the problems and vandalism was often cited as the biggest problem.

Yet this observation contrasts sharply with the fact that the level of criminal activity recorded in the UK has dropped significantly, or at least stabilised. Further to this, Innes asks why, in a period where it is generally accepted that there was a tangible decrease in victimisation, more people were concerned about physical and social disorder caused by young people? The answer is quite possibly to be found in the difference between formal criminal law, on the one hand, and the broad social conventions addressed by Spierenburg (2007) on the other: the social conventions are not covered by formal criminal law.

To formulate an answer to this question, a large-scale research in the UK was carried out within the framework of the National Reassurance Policing Programme (Tuffin, Morris & Poole, 2006).[1] From this research it appeared that people primarily attune their behaviour to suit their perception of potential risks. When it was asked in a general sense to classify the disorder phenomenon, the strongest signal was 'noisy, loitering young people', mostly because they 'are thought to vandalise property'. However, when the same question was posed to young people themselves, they identified 'other groups of young people' as the cause of the

1. Around 300 in-depth interviews were carried out involving persons living in 16 different areas of the UK. During the interview subjects were asked questions relating to their experiences with criminality, disorder, the police and social control in their own neighbourhoods as well as in other places.

disorder, in this case primarily through 'threats and violence'. Here, further analysis made it clear that the strong signal value of young people was first and foremost the product of a lack of homogeneity in their activities, and the high level of diversity characterising how they spend their time. Stated concisely, young people are read as being too unpredictable. In short, the implicit message here is 'be like everyone else and consume'.

4 Political geography strategies

Today, in urban geography we are seeing the inception of various political geography strategies which can be understood as a reaction to this dominant picture. With the term 'political geography', based on the notion of political economy as used by Marx,[2] we refer to the interference with public space, and the principles of supply and demand, on the one hand, and by state or other actors, on the other hand. Political geography points at the political goals that are strived for through mechanisms that seem to be self-regulating, but are in fact hidden. Without being exhaustive, we have made a distinction among four of these strategies, namely (1) criminalisation and punishment, privatisation, (3) dispersion and (4) exclusion. We successively deal with each of these strategies below.

Criminalisation and punishment

Here we are primarily thinking of the new regulations covering so-called 'Gemeentelijke Administratieve Sancties' (Municipal Administrative Sanctions), referred to as GAS in Belgium. These regulations essentially involve the formalisation of informal social control, at least insofar as this concerns the implementation of new punishments for specific forms of socially inappropriate behaviour. Although here we must mention that the law to a certain extent focuses on the administrative processing of acts which were already illegal, it also allows for punishment via the GAS regulations of all sorts of behaviour which could be considered as constituting or creating disorder. According to lawmakers, it had become necessary to penalise certain types of behaviours or phenomena which were previously not considered violations of the penal code (such as nuisances created by 'loitering young people').

What we are dealing with here are municipal by-laws, and in this sense this strategy is anything but new as Belgian law has for quite some time provided for the possibility to issue municipal police ordinances. However, it should be remembered here that these police by-laws (based on which action could be taken)

2. DE REUS, J., GORTER, H. (eds.) (1968), Manifesten, Klassieke teksten van Karl Marx, Voorwoord bij de Kritiek der Politieke Economie, Uitgeverij L.J.C. Boucher: Den Haag.

were in many cases so completely outdated that they could hardly ever be enforced (Van Heddeghem, 2006). Yet what is new in this context is the enforcement of such ordinances through administrative penalties. More specifically this involves a) levying administrative fines up to a maximum amount of EUR 250, b) the administrative suspension of an authorisation or permit issued by the municipality, c) the administrative cancellation of an authorisation or permit issued by the municipality and d) the temporary or permanent administrative closure of an establishment.

It is not the purpose of this paper to give a detailed account of the legal formalisation given (laboriously) to this strategy (Devroe, 2003), and for this we refer to other contributions which have in the meantime been published regarding this topic (Schuermans, 2005). Rather, in the first instance our goal here is to discover the logic behind this political geography strategy.

The crux of the matter concerns specific types of behaviour – which previously fell under the broad conventions applicable to public space described above and which were the exclusive remit of informal control – which are now being equated with other formally punishable acts through a formalised structure of criminalisation and punishment. In this sense it concerns a type of 'net-widening', i.e. a type of additional formal criminalisation and punishment.

The reporting of violations and transgressions which can be dealt with administratively via the GAS regulations is to a large extent handled by police staff, although other formal social control players are able to do this as well. In line with this, research has shown that in 2008 90% of reports were handled by police officers (VVSG, 2008). The number of police officers assigned to this type of task appears to vary from one municipality and city to another (Ponsaers & Vander Beken, 2008). As a result of this, a sort of parochial justice arises which is intended to facilitate a tailor-made approach that can be tied to a particular area.

Privatisation

A second strategy can be summarised as 'privatisation'. Over time, new, mixed (semi-private or semi-public) spaces have developed between so-called public and private spaces. Post-modern societies are increasingly establishing contracts for the regulation, control and supervision of spaces that are accessible to the public, but subject to certain conditions. In other words, these are private areas which can be accessed by the public. More than ever, the contractual terms in question are being decided by market conditions and not public authorities. As a result, individuals entering a football stadium, a dance club, a shop, a department store, a secure parking area, etc. are subjecting themselves to private, internal rules and no longer to laws applicable in public spaces.

In this context this means being subject to the rules in the private contract between owner and visitor. Or, more to the point, the space where a person is in fact becomes part of the consumptive behaviour, i.e. the opposite of what is traditionally considered public space. In short, if a person does not comply with the internal rules, possibly associated with paying an entrance fee, that person will not be granted access. Such a situation is described as 'policing by consent' (Stenning, 2000), or voluntarily complying with private contractual rules.

These mixed spaces, which are accessible by the public at large but subject to private rules, have their own private guards and have developed their own control and security measures (Jones & Newburn, 1999). Accepting these measures is part of the game and anyone refusing to comply with the plethora of private rules characterising these mixed spaces is not allowed inside. The formal guideline is clear: entering the space is a voluntary choice on the part of the visitor, who is not being forced to access said space. Once the person enters the space he or she agrees to the terms of the private contract. Increasingly, social inclusion or exclusion is being determined by the market and becoming a function of non-negotiated and unilaterally abstract private conventions. Informal social control is being replaced by stewards authorised to carry out body searches, doormen who check your identity, private security guards who record video images of people, etc. And those who cannot accept this situation simply have to stay outside. Privatisation is that simple. Apart from private policing agents, we can also give the example of the use of CCTV systems in public spaces. This type of continuous surveillance may very well lead to new types of mutual influences between private and public police forces (Francis, 2011).

The actual contents of the contract remain implicit and unknown to those wishing to enter, who have to submit to the discretionary behaviour of the private owner's representatives. The police's role in this context is becoming unclear because, as these mixed zones expand and the erosion of public space continues, the police retreat to the street, where they clearly have responsibility, and increasingly leave the handling of semi-private and semi-public spaces to the private sector. It is only when private regulators are confronted with problems that they cannot solve that they call on the police to use their legal monopoly on the use of violent force (Cools, Ponsaers, Verhage & Hoogenboom, 2005). Quite often now, private property owners sitting in their fortresses simply push a button to call the police at their own discretion, when they feel it is necessary.

This situation has led to the inception of a new form of reactivity. What Shearing and Stenning (1992) described as a partnership between public and private policing or a form of net widening should instead be understood as a shift in the balance of power. Security is no longer considered a political topic but rather a technical and economic phenomenon, and this means we can speak of the subjection of security to market mechanisms (Hoogenboom, 1994). Security is in fact being sold on the market at prices governed by the law of supply and demand.

Through the distribution of competences among the police (public roads, streets) and other players (private spaces), the private sector is enjoying increasing economic and quantitative dominance, or even supremacy. The more limited public space becomes, the greater the number of individuals who may be forced to submit to moral codes which regulate the private sphere. One question which arises here is if the private sector continues to push aside the police, how can the police then once again access this lost ground without violating the right to privacy? Also, how can the police act in a legitimate manner when the private contract continues to gain in importance to the detriment of the social contract between government and citizens? Moreover, how can the police protect citizens' rights when these same citizens freely submit to a situation whereby their rights are diminished? And finally, how can public space retain its essential political function, namely as a place where people can voice their opposition, when it continues to shrink and be smothered under the weight of regulations imposed within the private sector?

Dispersion

Certain specific political geography strategies are expressly aimed at dispersing, or in fact thinning out, so-called problem groups (read: young people in city centres). The traditional method used to implement a dispersion strategy is establishing a so-called restriction on public gatherings, which is in fact an administrative measure. In Belgium, restrictions on public gatherings are applied in a rather limited manner, but this is not the case in the Netherlands where the measure has been expanded and specifically targets young people who create disorder or commit crimes. The young people in question receive a letter clearly explaining that their behaviour will absolutely not be tolerated, and the person's parents are also informed (CCV, 2008). The police enforce the restriction and any violations are dealt with by the courts. The penalty for the first violation is either community work (if possible to be carried out in the person's neighbourhood) or a fine. Subsequent violations result in a much higher fine and there is also a possibility of imprisonment.

The dispersion strategy described above is a government measure and in that sense remains subject to some form of democratic control. Nevertheless, dispersion strategies are gradually becoming more and more prevalent in the private security industry while the authorities' role is kept rather passive. As an illustration of this point, one need only look at the progressive installation of devices known as 'Mosquito boxes'.

Mosquito boxes are devices that generate a sound with a very high ultrasonic frequency (over 8,000 hertz) which can only be heard by children and young people under 25 years of age. Anyone older is no longer able to hear this sound. Since 2005, this device has seen widespread use in the UK, and right from the beginning young people there reported that they suffered from ringing in the ears and headaches due to the Mosquito. So clearly this device targets groups of young people and is primarily used in shopping malls to break up and disperse groups of loitering teens. The past few years have seen a massive boom in the use of these devices in certain countries such as the UK, Germany, Switzerland and the Netherlands. The European Commission has said that it refuses to deal with this issue and has referred to member states' individual competency in deciding whether or not to prohibit this type of device.

It is estimated that there are already around 3,500 of these devices installed in the UK[3] in parks, shopping centres and city centres with the intention of preventing young people from forming groups. The Children's Commissioner for England has in the meantime set up a campaign regarding the fast-increasing use of Mosquito devices. In the Netherlands, due to a lack of any legal framework, around 300 Mosquitoes have been installed in 60 municipalities. Belgium also lacks any type of law regarding the Mosquito but the Minister for the Interior has issued an opinion saying that it is not necessary to draft a specific legal text since the device is already subject to the legislation on harmful products.

The first Mosquito device was installed in Belgium in 2008 near a bank branch in Aywaille, a town near Liège. However, a few months later the bank manager was asked by the municipality to remove the device, among other things after a complaint from a young person who suffered from headaches as a result of having to wait at a nearby bus stop (Belgian Senate, 2008).

During that same year, the manufacturer, Compound Security Systems, received around ten orders. In Belgium, the Mosquito led to a public debate about these youth-unfriendly devices and during that period the municipalities of Ixelles/Elsene, Antwerp, Turnhout and Zonhoven made clear their wish to prohibit the Mosquito on their territory. Moreover, the *Kinderrechtencommissariaat* (Flemish Children's Rights Commissioner) stated that it is unacceptable to treat young people in this manner and equated the use of these devices with an act of violence. In 2009 the Centre for Equal Opportunities and Opposition to Racism labelled the Mosquito a form of direct discrimination and called on municipalities

3. Council of Europe, Parliamentary Assembly, Doc. 12186, http://assembly.coe.int/Main.asp?link=/Documents/WorkingDocs/Doc10/EDOC12186.htm

to prohibit the device through police regulations.[4] Later that year Belgium rejected the use of the Mosquito device based on its incompatibility with the Convention on the Rights of the Child.

The situation in the Netherlands is more nuanced as the Ministry of the Interior and Kingdom Relations has devolved the decision-making authority for the placing of Mosquito devices to the municipal level. However, in June 2010 the Parliamentary Committee of the Council of Europe strongly urged that the device to be banned throughout Europe.[5]

Another form of dispersion involves the designation of specific locations in the city where young people are welcome to spend their free time. A typical example of this approach is the building of skateboarding areas or parks in certain cities in the US (Carr, 2010), and also in Flanders. Here we can see a dynamic of exclusion at work – skating by young people has after all been banned in a number of US cities – which at the same time creates the impression that alternative solutions are being offered and that a win-win situation arises as a result (Carr, 2010). However, these bans can lead to young people interpreting them as an effort to exclude their age group in general, at least as regards the city centre. In addition, according to Malone (2002) these initiatives do not always generate positive reviews as some of these parks are located on the outskirts of cities, and so can lead to practical (accessibility) problems in addition to feelings of exclusion. Moreover, the lack of social control in such isolated places can mean that young people do not always feel safe there (Malone, 2002). As such, designating places specifically intended for young people, even when this is well-intentioned, does not always have a positive effect, quite the contrary in fact.

Administrative exclusion

More than ever before, the Belgian government is also applying a political geography strategy aimed at geographical exclusion. For example, the past few years, more specifically those following the introduction of the so-called football law in 1998, have seen the inception of stadium banning orders.[6] With this law the government's intention was to bring football-related violence firmly under control. Among other things, in the law the government provided for the penalisation of (attempts at) gaining access to a football stadium or parts of it (enclosures, prohibited areas, etc.) after a prohibition order had been issued (whether

4. Centre for Equal Opportunities and Opposition to Racism, 22/1/2009, 'Mosquito' apparaat tegen jongeren http://www.diversiteit.be/index.php?action=artikel_detail&artikel=413
5. http://www.vng.nl/Documenten/vngdocumenten/2010_lbr/Ledenbrief_ECGR-U201000149.pdf
6. Law of 21 December 1998 on security during football competitions, Belgian Official Gazette of 3 February 1999.

verbally, via the stadium's internal regulations, etc., or for persons without a valid entrance ticket). A violation of these provisions can give rise to an administrative fine ranging from EUR 250 to EUR 5,000 and/or an administrative stadium banning order ranging from 3 months to 5 years. As far as minors are concerned, it is only possible to impose a stadium banning order.

In 2008, the Minister for State Enterprises, Mrs Inge Vervotte, made it known that she wanted to introduce administrative stadium banning orders for individuals or gangs who repeatedly display aggressive behaviour or create disorder in train stations.

The logic behind the prohibition policy is simple: 'People who cannot get into the stations also cannot get into the trains'. Yet in fact it has been possible to issue train station banning orders since 2004 and judges are allowed to do this for periods ranging from a minimum of 15 days to a maximum of 6 months as an additional measure, for example, in cases of assault. This legal measure has to date only been applied once, in 2006. At the beginning of 2010, Minister Turtelboom (Interior) repeated her support for this possibility (De Morgen, 10/2/2010) and made it known that she wished to grant such authority to the ticket inspectors of the public transport company in question (SNCB/NMBS).

Last year, a few Belgian cities and municipalities included in their police regulations the possibility of imposing a street banning order. This type of order prohibits specific persons from being physically present at certain locations during a specific periods of time. Street banning orders have been the topic of much discussion, both as regards appropriateness and legality but, despite this, more and more local administrations are showing interest in this administrative weapon in the fight against public disorder, primarily for matters relating to prostitution.

Belgian municipalities were inspired by what is being done in the Netherlands,[7] where street banning orders have been a tried and tested instrument within enforcement policy for large urban centres for several years. In the Netherlands, street banning orders are regulated via municipal police ordinances which introduce the legal arrangements for street banning orders within municipal law and also determine their general provisions (CCV, 2008). This measure was prompted by the need to be able to quickly and effectively deal with types of social disorder, for example, drug consumption in public, which can jeopardise the quality

7. Other European countries also have somewhat comparable police street banning injunctions (UK, Ireland), but in those legal systems it is more customary for a judge to issue such orders and not the municipal administration. Since among other things the competence of the administrative authority is one point of criticism levelled at the Belgian street banning order, the British and Irish examples are less relevant here.

of life in residential areas, primarily in major cities. Within municipal bodies, it is the mayor who has final responsibility for drawing up the execution orders.

The *Actieplan Overlast en Verloedering* (Action Plan on social disorder and deprivation) Social and Physical Disorder Action Plan), which was approved in 2008 by the Dutch Council of Ministers,[8] contains specific suggestions as regards so-called 'problem youths'. With this approach the intention is to break up group ties which are cause for concern, and with it mayors will have the authority to issue exclusion orders for groups of loitering young people for at most three months without needing to go through a judge (the so-called Serious Disorder bill). In cases of violations of exclusion orders, the law provides for a fine or a period of imprisonment. According to the law on youth welfare, municipalities are entitled to determine who is responsible for taking charge of matters if fast and adequate care is not provided. In addition, mayors are given the authority to, in extreme cases, force institutions to collaborate, have families who cause disorder evicted and removed from the neighbourhood, or assign a family coach. With this initiative, the government is going to create special centres for young people with problems without taking account of employment or housing issues. In fact, this boils down to mayors being able to evict people who cause disorder.

To answer the question of whether an administrative authority can also apply street banning orders as punishment in Belgium (and whether or not, for example, this is the exclusive prerogative of the judiciary), it must be stated that the mayor has in every way the possibility to impose orders and prohibitions in the framework of his duty to implement all police regulations (Dujardin, Somers, Van Summeren & Debyser, 1988). Moreover, legal doctrine emphasises that a mayor is entitled to take all measures against an individual or specific scope for the purpose of ensuring public safety, peace or health within the municipality (Mast, Dujardin, Van Damme & Vande Lanotte, 1988), as well as to prevent public disorder.

5 Conclusion: Does COP (Community-Oriented Policing) lead to public order police?

Now that we have had a detailed look at these four mechanisms we will take a closer look at the role played by the police. To a large extent, the body of ideas known as COP aims to de-formalise social control strategies in public spaces, among other things by focussing on the relationship between the general public and the police. However, in current Belgian security policy it is common to see

8. http://www.rijksoverheid.nl/documenten-en-publicaties/notas/2008/03/10/actieplan-overlast-e-n-verloedering.html

the opposite of this, i.e. informal frameworks being formalised instead of the other way around. In other words, often contradictions arise.

In addition, it is striking that, on the one hand, security policy is being influenced by private players yet, on the other hand, that it is also being presented as an administrative matter for the authorities. Insecurity issues are no longer problems related to criminality and its causes, but rather they are primarily problems related to public order. The problem of order in public spaces is no longer the exclusive competence of the police but also of a series of other watchmen, which does not preclude that the police are increasingly seen as the 'upholders of public order in society'. Although this notion may seem valid and worthwhile, there is a risk of overlooking certain negative consequences. This 'multi-level governance' (Squires, 2008) is seen as problematic.

In their well-known essay entitled *Broken Windows*, Wilson and Kelling (1982) suggested that it is the residents themselves who are best placed to define social order on their territory and determine how it can be achieved. After all, what is actually the suitable, correct or appropriate' level of this order? Wesley Skogan had already reflected on this in his *Disorder and Decline* (1990) and referred to it as a problematic notion.

1. First and foremost, we cannot do otherwise but accept that disruptions to public order take place in all communities. As such, there is a need to be realistic here, and anyone who believes that more police or watchmen can prevent this is lost in a utopian mindset.
2. Second, the question must be asked: Whose order are we actually talking about here? Some are inclined to say that maintaining the existing balance (of power) in society is one of the basic tasks of police services and that in that sense policing should be brought back to its core function of protecting the 'reasons of state' and acting as the praetorian guard of the political and economic regime. Others tend to say that the police can only exist because there is a community, and the adage for the old crime-fighting model 'What would the police do if there was no law?' is in this case changed to 'What would the police do if there was no community?' In short: the community uses the police as an instrument and not the other way around.
3. However, those who prefer such a community-oriented policing model hold third place, with an underlying paradox. After all, the question arises once again: Whose order are we talking about? In those types of communities which are relatively homogeneous, there is perhaps a shared standard, a shared expectation as regards a sort of preferred order. But what about the increasingly heterogeneous, late modern communities which, for quite some time now, have not had a territorial link with a specific geographical location? Some, like Skogan (1990) and Wilson & Kelling (1982), posit that order in het-

erogeneous neighbourhoods does not occur naturally and that it has to be negotiated. This of course raises many questions.

Who speaks in whose name? Who represents whom? Who is then responsible for controlling them? Insofar as society becomes more diverse, is there not also increased diversity in the types of orders desired? Which majority's order are we talking about? Do minorities and young people not have a right to a type of order which is right for them? Is there not a risk that the police will increasingly focus on complaints made by cranky loudmouths while reticent (more docile and tolerant) groups in the community are overlooked?

4. In fourth and final place there is a completely different consideration. In fact, the notion of public order is essentially a political concept and, much more than the concept of criminality: it depends on the political context at a given time based on which it is given a flexible interpretation. Whether a street demonstration is allowed, a police order issued to prohibit loitering teens from accessing certain parts of the city, an advertising campaign authorised in the city centre, efforts made to clean a polluted river, drunkards or other groups such as senior citizens allowed to constantly occupy park benches, caravan dwellers authorised or not use the strip of land next to the freeway, etc., all involve political decision-making (by the executing authority) and for each situation there is a need to decide whether or not it qualifies as a public order problem.

   In this context, 'disorder policing' may be given extremely broad discretionary powers as a result of which administrative authorities are presented with a bottomless well of possibilities (and power). As far as public servants work in the name of the common good, this power need not be the source of problems, but the risk of abuse and arbitrary application is lurking right around the corner, if only because considerations of individual liberty receive less attention. One cause for concern is that applying guidelines, as described by Kelling and Coles (1996), to limit and monitor the police's discretionary powers is insufficient. This would certainly be the case in villages and cities where far-right political ideologies are gaining influence and votes, and here there is certainly a need to acknowledge the danger and address it in public discussions.

References

Belgian Senate, session of 2007-2008, 28 May 2008, Proposal for a resolution regarding the introduction of a ban on the use of 'Mosquito' type systems to drive away loitering teens.

Carr. J. (2010). Legal Geographies – Skating around the edges of the law: urban skateboarding and the role of law in determining young peoples' place in the city. *Urban Geography*, *31*(7), 988-1003.

CCV (2008). Aanpak hangjongeren in gemeenten, http://www.hetccv.nl/binaries/content/assets/ccv/webwinkel/aanpak-hangjongeren.pdf (retrieved on 10/5/2012)

Commissie Kleine Criminaliteit (1986). (Roethof Commission) Interim report 1984. Final report.

Cools, M., Ponsaers, P., Verhage, A., & Hoogenboom, B. (2005). *De andere Rechtsorde. Demoncpolisering van fraude-onderzoek.* Brussels: Politeia.

Crawford, A. (2006). Fixing Broken Promises? Neighbourhood Wardens and Social Capital. *Urban Studies, 43*(5/6). 957-976

(2010), Turtelboom denkt aan stationsverbod. De Morgen, 10 February 2010,

Devroe, E. (2003). Last van overlast: volle G.A.S. vooruit? in: "Last van overlast", *Orde van de dag*, Kluwer, afl. 24, 15-17..

Dujardin, J., Somers, W., Van Summeren, L. & Debyser, J. (1988). *Praktisch handboek voor het gemeenterecht.* Brugge: Die Keure.

Durkheim, E. (1947). *The Division of Labor in Society*. Glencoe, IL: Free Press.

Ellen, I. G. & O'Regan K. (2010). Crime and urban flight revisited: The effect of the 1990s drop in crime on cities. *Journal of Urban Economics 68*(3), 247-259.

Fielding, N. & Innes M. (2006). Reassurance Policing, Community Policing and Measuring Police Performance. *Policing and Society*, *16*(2),127-145.

Francis, V. (2011). *Caméras et sécurisation des espaces de la Société Nationale des Chemins de fer Belges: uni mise à l'épreuve des hypothèses relative aux transformations de l'art de conduire les conduites dans les sociétés libérales avancées.* Unpublished thesis: Louvain-la-Neuve.

Goeminne, B., Enhus, E., & Ponsaers, P. (2003). *Criminaliteit in de publieke ruimte – Een inleiding in de hotspot analyse*, Brussels: Reeks Dienst voor het Strafrechtelijke Beleid..

Hoogenboom, B. (1994). *Het politiecomplex; over de samenwerking tussen reguliere politie, bijzondere opsporingsdiensten en particuliere recherche*. Arnhem: Gouda Quint.

Innes, M. (2005) *'Why Disorder Matters? Antisocial Behaviour and Incivility as Signals of Risk'*, Scarr Conference.

Jones, T., & Newburn, T. (1999). Urban Change and Policing: Mass Private Property Re-considered. *European Journal on Criminal Policy and Research* 7, 2, 225-244.

Jones, T., & Newburn, T. (2002). 'The Transformation of Policing? Understanding Current Trends in Policing Systems', *British Journal of Criminology*, *1*, 129-146.

Kelling, G. & Coles, C., (1996), *Fixing Broken Windows: Restoring Order and Reducing Crime in Our Communities*, New York, Free Press.

Lofland, J., Snow, D., Anderson, L., Lofland, H., (2006). *Analyzing Social Settings*. Belmont,Wadsworth Publishing Company.

Maciocco, G. & Serreli S. (eds) (2009). Enhancing the City: New Perspectives for Tourism and Leisure-Dordrecht, Springer Netherlands.

Malone, K. (2002). Street life: youth, culture and competing uses of public space, *Environment and Urbanization*, *14*(2) 157-168.

Mast, A., Dujardin, J., Van Damme, M., & Vande Lanotte, J. (2002). *Overzicht van het Belgisch Administratief Recht.* Mechelen: Kluwer.

Mitchell, D. (1995). 'The End of Public Space? People's Park, Definitions of the Public, and Democracy.' *Annals of the Association of American Geographers 85*(1), 108-133.

Pleysier, S. & J. Deklerck (2006). Over hondenpoep en hangjongeren. Een verkennend onderzoek naar overlastfenomenen in parken en groenzones. *Tijdschrift voor Veiligheid*, *5*(1) 5-20.

Ponsaers, P., & Vander Beken, T. (2008). 'Onderzoek naar de implementatie van de GAS-wet in België: cijfers en discussies', In *Update in de Criminologie* (nr.4), Overlast en de maatschappelijke aanpak ervan. Mechelen: Kluwer.

Schuermans, F. (2005). 'Gemeentelijke administratieve sancties en gewapend bestuursrecht: naar een embryo van bestuurlijke parketten', In *Gewapend Bestuur*, Cahiers Tegenspraak, Brugge, Die Keure, 21-66.

Shearing, C. (1992). 'The relation between Public and Private Policing'. In M. Tonry. & N. Morris (Eds.),*Modern Policing, Chicago-London,* Chicago, The University of Chicago Press. Vol. 15, pp. 399-434.

Skogan, W. (1990). *Disorder and Decline: Crime and the Spiral of Decay in American Cities*. New York: Free Press.

Spierenburg, P. (2007). 'Apegatjes achtervolgers – Sociale controle tussen verschuivende geweldsdrempels en veranderende noties van privacy'. In H. Van de Bunt, P. Spierenburg, & R. van Swaaningen, (Eds.), *Drie perspectieven op sociale controle*, Den Haag: Boom Juridische Uitgevers, 67-112.

Squires, P., Ed. (2008). *ASBO Nation: The Criminalisation of Nuisance*. Bristol, the Policy Press.

Stenning, P. C. (2000). 'Powers and Accountability of Private Police.' *European Journal on Criminal Policy and Research 8*(3), 325-352.

Tönnies, F. (1935) *Gemeinschaft und Gesellschaft: Grundbegriffen der reinen Soziologie*. Darmstadt: Wissenschaftliche Buchgesellschaft.

Tuffin, R., Morris, J., Poole, A. (2006). *An evaluation of the impact of the National Reassurance Policing Programme,* Home Office Research Study 296, Home Office Research, Development and Statistics Directorate. http://www.homeoffice.gov.uk/publications/science-research-statistics/research-statistics/police-research/hors296/hors296?view=Binary (retrieved on 10/5/2012)

Van den Herrewegen, E. (2008). 'De stad als een belevenis. Bedenkingen bij onderzoek naar overlast en stedelijkheid'; In *Update in de Criminologie* (nr.4), Overlast en de maatschappelijke aanpak ervan. Mechelen: Kluwer, 171-190.

Van Heddeghem, K. (2006). 'Bestuurlijke aanpak van overlast: de gemeentelijke administratieve sancties', *Panopticon*, 2, 30-31.VVSG (2008). Sa*menvatting Onderzoeksresultaten Onderzoek Gemeentelijke Administratieve Sancties*.http://www.vvsg.be/veiligheid/bestuurlijke%20handhaving/GAS/Documents/VVSG%20resultaten%20studie%20GAS.pdf (Retrieved on 10/5/2012)

Wilson, J. and Kelling G. (1982). Broken Windows: The Police and Neighbourhood Safety. *The Atlantic Monthly*, 249, 3, 29-38.

Wood, M. (2004). *Perceptions and Experience of Antisocial Behaviour: Findings from the 2003/2004 British Crime Survey*. Home Office Online Report 49/04. http://library.npia.police.uk/docs/hordsolr/rdsolr4904.pdf. (Retrieved on 10/5/2012)

Police and surveillance in Paris: are the French Police becoming knowledge workers and risk managers?

*Anaïk Purenne**

1 Introduction

Data collection and analysis are not a recent phenomenon. However, major changes have occurred in recent decades. Surveillance now appears as a general tool in modern institutions and as a symptom of a 'softening' of social control, especially in the field of security and crime control (Marx, 1988; Lyon, 2007). This idea of 'softening' describes the extension of new means or 'tools of government' (Hood, 1983), which are less focused on direct and coercive treatment than in the past. These softer forms of social control do not appear as less threatening than the use of force: on the contrary, the academic discourse is clearly a critical one.[1] This trend is well-documented in the context of North America, for example, in the police institution; examining another local context, the French police, will be the main objective of this chapter.[2]

As in other European countries, surveillance is now an important issue in the French political arena. For instance, many official speeches stress the need to modernise the police through surveillance technologies-based innovations. During the past few years, successive home ministers have indicated their continued determination to carry on with the development of databases (national and genetic fingerprints files, etc.), to develop a '"mass" scientific police for purposes

* Anaïk Purenne is a researcher in sociology at the University of Lyon (joint unit CNRS UMR 5600 Environnement Ville Société/ENTPE). She is currently involved in the UNESCO Chair "Urban Policies and Citizenship" and is also part of the Living in Surveillance Societies (LiSS) COST programme. Her research interests focus on police innovations, surveillance, new managerialism, and community involvement in France and in Canada.

1. For instance, Gary Marx writes: 'the first task of a society that would have liberty and privacy is to guard against the misuse of physical coercion on the part of the state and private parties. The second task is to guard against the "softer forms" of secret and manipulative control. Because these are often subtle, indirect, invisible, diffuse, deceptive, and shrouded in benign justifications, this is clearly the more difficult task' (Marx, 1988, p. 233).
2. The present text is based on a paper presented at the XVII International Sociological Association Congress, (Gothenburg, Sweden, July, 11-17, 2010), in a session about 'local manifestations of global surveillance' organised by Torin Monahan and David Lyon. I am deeply grateful to Urvashi Singh for her supportive help in writing this text in English, and to Anne Wuilleumier for our stimulating exchanges.

other than investigations of serious crimes'[3] (a workshop initiated since 1995), to extend 'crime analysis tools to the fight against petty crime',[4] or to set up new computerised databases like 'gang related files that allow investigators to identify and arrest gang members who cause trouble'.[5] This emphasis on new methods and uses of information coincides with a moment of institutional crisis. Indeed, a new concern about the State and the legitimate use of violence has emerged in social discourse in the past few years. Stating an example in the current French context: the use of force (stops and search, identity checks, arrests, admissions in police custody, etc.) provokes more intense protest than in the past, especially in the poorest suburbs (Wuilleumier & Purenne, 2011).

The question is whether these initiatives are rhetorical or reality, even if Jérôme Ferret and Vincent Spenlehauer underline, following Michel Foucault's warning, that 'surveillance projects ... are destined either to remain projects or to take shape only partially' (Ferret & Spenlehauer, 2008, p. 11). Keeping in mind this idea, our concern is to examine to what extent these political initiatives around surveillance are beginning to take shape within the French police and to what extent they contribute to a significant change in police practices.

This analysis is based on an empirical study relating to two different services belonging to the Paris police department: an urban police service, on the one hand, and a transport police service patrolling the Paris metro and rail network on the other. The local experiments I studied are limited and not fully representative of the national French police force. Indeed, the aim was to focus on services dedicated to experiment with new methods (like prototypes), rather than representative examples of the French context. In that perspective, the Paris police department appears as a relevant case, since it is historically conceived as the first 'laboratory of modernity' (Berlière, 1996) within the French 'Police Nationale'.[6] The two cases I studied illustrate this trend and have experimented with important

3. Michèle Alliot-Marie, Home Minister, June 10, 2009, http://www.interieur.gouv.fr/sections/le_-ministre/interventions/archives-mam/colloque-as-securite/view
4. Id.
5. Brice Hortefeux, Home Minister, November 12, 2009, http://www.interieur.gouv.fr/sections/le_-ministre/interventions?b_start:int=20
6. The 'Police Nationale' is the main civil law enforcement agency, with jurisdiction in urban areas and large towns. The 'Gendarmerie Nationale' is a military institution in charge of public safety, with jurisdiction in rural areas and small towns. Since 2009, both have been placed under the jurisdiction of the Home Ministry. Both institutions are highly centralised and characterised by hierarchical, top-down modes of management. Within this standardised administrative system, the Paris police department which is part of the 'Police Nationale' has a unique position. Indeed, the 'préfet de police' (police commissioner) at its head is directly accountable to the home minister and has a larger autonomy to experiment with new projects and methods which can then permeate the whole police force (Renaudie, 2008). Stating an example from the last decade: the importation of Compstat, a new tool of management developed by the New York City Police Department, was first tested in 2001 by the Paris police commissioner, before being used in 2002 as a new model of management for all police and gendarmerie services in France (Didier, 2011).

reforms in the last decade. More particularly, both examples exemplify the rise of surveillance (i.e. data collection and analysis) and knowledge work dimension.[7] In the first part of the chapter, I examine the paradigm developed by Ericson and Haggerty (1997), who defend the idea that the contemporary police have become 'knowledge workers', devoted to the anticipation of various kinds of risks. Taking into account the development of surveillance and the transition to a 'risk-thinking society', this paradigm provides a useful theoretical framework and concepts to examine the potential transformation of police work. The second part of the chapter analyses the example of the Paris urban police service, where uses of information appear to be quite convergent with the framework proposed by Ericson and Haggerty. Indeed, police officers are pushed by their hierarchy to collect information about risky places or activities and to communicate with vulnerable social groups in order to minimise crime opportunities. However, this growing concern for crime prevention and knowledge working does not exclude the use of more traditional means and logics of crime control. The case study of the transit police service also confirms this idea. It suggests that surveillance technologies can be used to target or 'rationalise' the use of force. By the term 'rationalise', I mean that the places identified as high-risk are placed under intensive police scrutiny to create a situation in which coercion becomes highly visible. Nevertheless, as Peter Manning (2008) has argued in his analysis on the effects of crime mapping, this dominant use of surveillance technologies sustains a very conservative approach to policing. Consequently, these French experiments document the idea of a limited diversification of the police toolkit, rather than the hypothesis of a radical transformation of police work.

2 The police as risk managers: a new conceptualisation of police work

For many scholars of surveillance, the spread of risk-thinking (Beck, 1992) and risk management appears as 'an increasingly important spur to surveillance' (Lyon, 2003, p. 8). According to David Murakami Wood and William Webster, 'the new technologies are placed in the service of this agenda: collecting and sorting data on people, things and events in order to produce categories of risk …. As the "risk-surveillance society" has become the "ideal-type" state of the twenty-first century, so its aims – anticipated and premanaged risk, safety, control, security – are increasingly permeating policy and practice at every level' (Wood & Webster, 2009, p. 3-4).

7. It must be noted that these services have been studied in two different research projects. Nevertheless, the methodology and the theoretical background remain quite similar. Moreover, the analysis is not strictly a comparative approach, but a broad overview of our case studies in order to understand if convergent or divergent trends are at stake regarding the effects of surveillance tools.

For Ericson and Haggerty, this trend is especially salient in the field of security and crime control and the police play a pivotal role in this 'risk society'. Following Foucault (1977) and Beck's (1992) ideas, Ericson and Haggerty propose a new conceptualisation of police work: the police must be viewed as knowledge workers and 'centers of calculation' (Latour, 1987) of various kinds of risks, rather than as law enforcement agencies (Ericson & Haggerty, 1997). From various examples within the Canadian context, Ericson and Haggerty argue that information gathering and risk communication have become time-consuming activities and even central tasks for police officers. For instance, in the largest police departments, dedicated units are in charge of gathering information and processing knowledge on risk.

This growing concern for risk management and crime prevention is not only an internal affair but an inter-institutional concern. In fact, the information gathered by the police is used by a wide range of public and private organisations. So the police have become 'information brokers to institutions [which] operate based on a knowledge of risk'. The knowledge supplied by the police is used by these institutions to implement risk minimisation and actuarial strategies. The insurance industry offers a good illustration of this idea. Insurance companies depend on the first-hand information gathered by police officers in cases of traffic accidents or burglaries. These data are used by the companies to develop their own protocol of risk-profiling, which eventually leads to an increase in insurance rates for high-risk drivers or to encourage victimised shopkeepers to install specific equipment to reduce crime opportunities.

Moreover, the police also use this knowledge in order to prevent crime and to 'provide [local groups] with expert advice and ... help them manage their risks' (Ericson & Haggerty, 1997). Ericson and Haggerty analyse the community policing initiatives that are routinely developed by many police departments as a mere discourse of rationalisation of this risk management approach. These initiatives currently involve what David Garland calls responsibilisation strategies. For Garland, this strategy 'involves the central government seeking to act upon crime not in a direct fashion through state agencies ... but instead by acting indirectly, seeking to activate action on the part of non-state agencies and organizations. ... Its primary concern is to devolve responsibility for crime prevention on to agencies, organisations and individuals which are quite outside the state and to persuade them to act appropriately' (Garland, 1996, p. 452). In this perspective, the core activity of the police is not direct control over the neighbourhood,[8] but a more distant and insidious form of control through information campaigns targeting local communities and 'at risk' groups.

8. 'The traditional community of interpersonal relations and direct action gets lost' (Ericson & Haggerty, 1997, p. 445).

3 The Paris urban police service as information brokers

The Paris Police Prefecture has a unique position within the French 'Police Nationale' and it has the status of experimental laboratory (cf. supra footnote 6). For instance, at the end of the 1990s, the left-wing government of Lionel Jospin and the Home Minister Jean-Pierre Chevènement decided to develop a kind of community policing ('police de proximité') model within the French national police. The urban police service, which is one of the largest services of the Paris Police Prefecture, has been designated to experiment with new practices before generalising the 'police de proximité' model (Roché, 2005). This urban police service is composed of twenty police stations, one per '*arrondissement*' or district.[9] Since the implementation of the 'police de proximité' reform in 1999, the urban police service ensures the security in city streets. To fulfil this mission, the police stations are requested to develop a special partnership with the elected officials and the population. Through this networking, each police station centralises information about the problems faced by different social groups and help them minimise crime opportunities (Purenne, Aust & Ferret, 2006). In that perspective, the police could be viewed as knowledge brokers to local organisations or actors. But my interviews and observations also show that police officers are still involved in direct control of public spaces and law enforcement activities.

The police station: a node of risk communications

The empirical study conducted in three Paris police stations confirms the idea that a 'core technology of police work is information'.[10] In this framework, the police officers and their hierarchy develop relations with local actors and organisations. The purpose of this networking is to produce knowledge of risk that could be useful to anticipate problems.[11]

This elaborate networking is based on a kind of 'division of relational labour'. The chiefs of the police station have a central position in these relational activ-

9. In September 2009, the competence of the urban police service was extended beyond the limits of Paris city to three administrative divisions around Paris: the *Hauts-de-Seine*, the *Seine-Saint-Denis*, and the *Val-de-Marne*. The former police services charged with the security in these areas were merged with the services of the Paris Police Prefecture.
10. Interview with A., former chief of an arrondissement police station, 26 April 2005.
11. This case study was part of a comparative study in four cities: Paris, Lyon, Montreal and Barcelona. The initial purpose of this study, financed by the French Ministry of Research, was to understand how a new policy initiated by the political authorities could permeate police organisations whose professional interests were supposedly hostile to this policy. In Montreal and Paris, our data highlighted the importance of networking between police stations and their local environment, which contributed to define local agendas for police action. The empirical study is based on interviews with management staff, investigators, police officers and direct observation of their activities. The fieldwork in Paris took place between August 2005 and February 2006. I have updated this material by reading official documents published since then.

ities. They make sure that they develop personal relations with all actors concerned with tackling crime: elected officials, community associations, social services, school authorities, local shopkeepers, etc. Within this network, they generally have 'a special collaboration'[12] with the district mayor. Indeed, the mayor is himself at the centre of a broad network of communications and is generally well informed of local problems. Since the reform of 1999, a parallel circuit of information collection has also been set up in each police station. Community police officers are in charge of developing close relations with local residents, caretakers and occupations exposed to crime such as pharmacists, café owners or grocers. During their routine patrol, their duty is to collect information and complaints about the problems faced by vulnerable shopkeepers and inhabitants, and to report them to their hierarchy. In addition, specialised units created in 1999, the 'prevention and communication units', organise meetings with senior residents or school authorities to detect specific problems like thefts targeting elderly people or the rise of misbehaviour at school. They also inspect places or occupations identified as risky (like jewellers, banks, etc.) to detect at-risk situations. Another unit placed under the command of a police lieutenant, the 'bureau of operations', is the specific interlocutor both for external institutions (such as roadway authorities and public transport network) and for other police services within the Paris Police Prefecture.

Throughout this elaborate communications circuit between the police and external institutions and individuals, the chief of each police station generally has, on a day-to-day basis, an in-depth and comprehensive knowledge of local problems and security needs of the various groups.

A set of tactics including responsibilisation

These data, information, complaints and security needs are used to identify populations or places that are exposed to various kinds of troubles (theft, racketeering, noise pollution, etc.), and to diagnose the sources of the problem in order to provide appropriate solutions. This work of diagnosis is based on a qualitative analysis of the information collected by police officers, and/or on statistical analysis distributed by analysts at the headquarters, for instance, when a specific problem goes beyond the boundaries of a single district. The police lieutenant placed at the head of the bureau of operations has a central role in this process of risk analysis and management. This lieutenant is generally a senior officer with relational skills and has the complete trust of the chief of the police station.

Once a problem and its causes have been identified, different kinds of solutions can be implemented, depending on the nature of these problems. At first, special operations of law enforcement can be set up at a 'hot spot'. Police officers at the local police station or from other units are requested to target this zone and to

12. Interview with M., chief of a police station, 12 October 2005.

issue summonses to minimise the trouble. Once the operation is completed, an evaluation is communicated to local actors to show that the police are fully in charge. According to the management staff I interviewed, this 'hot spot policing' strategy is a major part of their units' workload, besides responding to emergency calls.[13]

> 'In analysing local crime in great detail, each police station deploys their officers in operations targeted at very specific objectives (crimes and offences committed in specific places and at specific times: street violence, drug trafficking, peddling, etc). They receive the support of specialised units with powers that extend across the entire capital or to specific parts of it, such as the Security Squad and Anti-Crime Units.' http://www.prefecturedepolice.interieur.gouv.fr/Pied-de-page/English/Police-missions/The-urban-neighbourhood-police-department

Another kind of tactics that echoes the responsibilisation strategy identified by D. Garland is to associate external local institutions and to give them expert advice to minimise troubles. For instance, a lieutenant I interviewed explained that the solution for the problem of road accidents 'can be to intensify police presence, to issue summonses, but also to alert the Paris City Department of Transportation and propose a reengineering of the road or new road signs'.[14] Diverse 'techniques of persuasion' are also implemented 'to anticipate the problems' or 'resolve the problem at its source'.[15] For instance, information campaigns are organised to sensitise the most vulnerable groups and to persuade them to change their habits in order to limit criminal opportunities. This role is especially designated to the police officers of the prevention and communication units.

> 'Informing, alerting and raising awareness: the [urban police service] engage in wide-ranging crime-prevention work to protect all section of society, particularly the most vulnerable. The prevention and communication units of the arrondissement police stations organise talks for elderly people (advising them how to take care on the streets, guarding against bogus callers, etc), children and teenagers (the dangers of drugs, racketeering and all forms of violence in the school, road safety), tourists and people in occupations where they may be exposed to crime.' http://www.prefecturedepolice.interieur.gouv.fr/Pied-de-page/English/Police-missions/The-urban-neighbourhood-police-department

13. The twenty police stations receive nearly one million calls per year via the police emergency line http://www.prefecturedepolice.interieur.gouv.fr/Pied-de-page/English/Police-missions/The-urban-neighbourhood-police-department
14. Interview with J., senior officer, Bureau of Operations at a police station, 9 November 2005
15. These expressions are frequently used by police officers.

These campaigns take different forms. Brochures can be distributed that list habits to avoid, for instance, for children exposed to racketeering: 'take little cash with you', 'do not stay isolated on the street'; or for tourists: 'do not leave your methods of payment in your coat on a chair (at restaurant or in cafés for instance), a pickpocket could sit behind you and steal your belongings'.[16] Conferences and debates are also organised by police officers to meet specific groups. In addition, these officers set up training sessions for representatives of vulnerable occupations who in turn can then develop their own awareness campaigns. At the same time, home surveillance is organised by police stations during summer holidays to prevent burglaries. The variety of these activities is shown in this table.

Vulnerable groups	**Category of risk**	**Kind of responses**
Youths	Use of drug Racketeering Misbehaviour and violence at school Road safety	School programmes : conferences, debates, visits 'Concerts citoyens' initiative
Seniors	Road safety Burglary Theft Violent theft	Conferences, debates, visits Meetings with external institutions (social services, etc.)
Shopkeepers (banks, jewellers, pharmacists…)	Burglary Theft	Meetings with partners Advices and information about security equipment (alarms, CCTV, etc.) *'Tranquillité vacances'* initiative (routine patrols and shop surveillance)
Property owners	Burglary Noise pollution Damage	Meetings with partners Inspection and expert advices (security equipment, etc.)
Residents	Burglary	*'Tranquillité Vacances'* initiative (home surveillance)

Although they may seem discreet, these activities of persuasion represent a substantial amount of time, both in the urban police service and in other units of the Paris Police Prefecture. For instance, an internal document indicates that from 2008 to 2009, nearly 3,000 conferences were set up in schools and that police officers met 75,000 children.[17]

This strategy of responsibilisation has become a routine and institutionalised one in Paris, even after the 'police de proximité' model lost legitimacy through politi-

16. http://www.prefecturedepolice.interieur.gouv.fr/Pied-de-page/English/Publications/Paris-in-complete-safety
17. Préfecture de police de Paris, Bilan d'activité. 2009.

cal rhetoric and was supplanted by more aggressive discourses (Monjardet, 2006). Indeed, this diversification of roles and strategies reinforces the power and the influence of the police. The more the police take into account the wide-ranging demands of citizens, the more pivotal they become and the more legitimacy they gain. Moreover, the responsibilisation strategy is not irreconcilable or diametrically opposed to more traditional tactics, like saturation patrols or hot-spot policing. On the contrary, such a diversification is viewed by police managers as an extension of the police toolkit. These practices aiming to convince potential victims to change their habits coexist with traditional tactics based on means of coercion against offenders (summonses, arrests, admissions in police custody, etc.). These observations suggest a kind of 'policy mix' rather than the radical change suggested by Ericson and Haggerty.

4 The Paris transport police service: a highly visible use of force in high-risk territories

As mentioned above, a new political rhetoric that is dominated by the idea of engaging 'war against crime' has developed in France over the last decade. This new discourse takes place in a context of continuous rise of crime rates, which was a major issue during the electoral campaign in 2002 (Monjardet, 2006). After the presidential elections and the arrival of a right-wing government, the new home minister at that time, Nicolas Sarkozy, initiated several workshops ('new management' of security, increasing staff and material means, technology-based surveillance innovations, etc.) to sustain the impression that police can win this 'war against crime'.

The Paris transport police service is one of these measures. Paris city was confronted with high rate of crime and delinquency in the public transport system. In order to address this problem, which has negative repercussions for the image of the capital city, the home minister announced the creation, within the Paris Police Prefecture, of a service devoted to ensuring the security of public transport users throughout Paris and its suburbs (approximately 7 million passengers each day).[18] New methods and new uses of information technology (CCTV, crime mapping, databases, etc.) were promoted to 'eradicate not only the feeling of insecurity, but insecurity itself'.[19]

18. Préfecture de police de Paris, Le Service Régional de Police des Transports de la Préfecture de Police, avril 2005.
19. Interview with Nicolas Sarkozy published by *Le Figaro*, 19 May 2002.

A reorganisation based on a new division of labour and the use of surveillance technologies

The Paris transport police, whose creation has been popularised through the media, are not an entirely novel entity.[20] Indeed, two services in charge of security in the public transport system previously existed. The first one, created in 1976, had jurisdiction in the Paris metro; the second one was devoted to regional and local trains (RER) outside Paris. The first service was based largely on the experience and discretion of police officers. Due to hierarchical pressure to issue summonses and make arrests, these police officers often developed practices like *'contrôles au faciès'* (racial profiling) (Paperman, 1992). In 2002, these two services were merged in a unique service with regional authority, and new surveillance devices such as CCTV, databases, and crime mapping, were developed to allow a 'rational' targeting of the use of force.

The staff is divided into two main divisions: a patrol division organised into 160 police squads, and a criminal investigation division organised into four specialised units. Within the patrol division, a team of computer experts and intelligence analysts has been set up in order to produce knowledge about crime in public transportation. One of the main tools of these experts is the creation of a computer database containing detailed and exhaustive information about each crime committed in public transport: date, time, location, type of crime or injury, kind of modus operandi, distinguishing features of the offender, etc. Coupled with crime mapping tools and with an extensive video surveillance system of 9,500 CCTV covering the whole regional rail network, it permits the team of analysts to develop risk-profiling activities and to identify trends and patterns of crime across time and space.[21] This knowledge of risk is controlled and appropriated by the top management. It gives them the possibility to reduce the autonomy of police officers and practices of racial profiling that were widespread in the former transport police.

A paradoxical crime-attack strategy: a highly visible use of force

In this new division of labour, the discretion of police officers is strongly diminished. Officers of the 160 squads of the patrol division are simply requested to

20. This case study was part of a comparative study of four transport police services in France and in England. This comparative study was funded by the *Institut National des Hautes Etudes de Sécurité*. The initial aim was to understand how police forces interact (or not) and share information with other actors, such as public transport authorities, rail companies, private security agencies, etc. (Hamelin, Aust, Purenne, & Spenlehauer, 2007). Given the centrality of police forces on this issue, the study finally focused on them. In the case of the Paris transport police, the data was initially collected in May-June 2007 through interviews. It has then been updated through the reading of official speeches, policy guidelines and agency reports.

21. Préfecture de police de Paris, Le Service Régional de Police des Transports. op.cit.

'be in the right place at the right time'. At first, routine police patrols are deployed in crime hotspots identified through risk-profiling activities. By their permanent presence on the favourite territories of offenders committing thefts, violent thefts, and assaults (the most frequent crimes in Paris transport area), they are intended to deter crime before it is committed. As the official presentation of the Paris Transport Police explains, 'In response to the random and widespread nature of the crime on public transport, a division has been set up. … Every day it deploys … patrols at 600 railway stations and on 700 … trains, focusing its efforts on the routes and the times of day that pose the greatest problems.'[22]

At the same time, large-scale operations, bringing together several hundred police officers from several departments, are organised each week at a given place and time to create a situation in which coercion, or at least the threat of coercion, becomes highly visible in order to discourage potential offenders. One of the specific characters of the Paris context is that the Paris police commissioner (or 'préfet de police') has authority over a wide range of external services, including local units of the National Police and of the National Gendarmerie in the whole Île-de-France region, but also over private security forces hired by the railway companies. This huge pool of force is clearly a crucial resource in developing this kind of crime control strategy based on means of coercion against offenders, even if these means are only partially used within the transport police.

Indeed, a remarkable feature of this crime-attack strategy is that this massive deployment of force remains essentially ostentatious, in the sense that identity checks are promoted by the hierarchy to the detriment of arrests and admissions to police custody. To limit the autonomy of the patrol squads, the top and middle management impose a maximum number of arrests. Indeed, arrests take time and generate a lot of paperwork. As a commanding officer of the patrol division explains, 'There is no quota in this service. The main purpose is to show ourselves. The guys are required to be in the right place at the right time.' This rule does not imply that police officers remain inactive when confronted by offences: they are encouraged to give a simple warning or to check the identity of the offenders, rather than to criminalise the incident and prosecute them. This management and the use of surveillance devices contribute to a 'rational' targeting of the use of force that sustains 'the myth that the sovereign state is capable of providing security' (Garland, 1996, p. 448).

> 'Following the creation of the Regional Transport Police Service in September 2002, crime on the region's rail networks has dropped by nearly half, with a particularly significant drop in pick pocketing being recorded.' http://www.prefecturedepolice.interieur.gouv.fr/Pied-de-page/English/Police-missions/The-urban-neighbourhood-police-department

22. http://www.prefecturedepolice.interieur.gouv.fr/Pied-de-page/English/Police-missions/The-urban-neighbourhood-police-department

Presented as a successful device, this local experience has been rapidly institutionalised and extended through the creation of a 'Service National de Police des Transports'[23] in June 2006. Directed by the former chief of the Paris transport police who was promoted at the head of this new National Service of Transport Police, it is based on similar methods and strategies to provide security on the national rail network. Furthermore, in April 2010, after a series of assaults against ticket inspectors in the Paris suburbs, a new national programme for security in public transportation ('Plan national de sécurisation des transports') was elaborated by the Home Ministry. Rather than taking the hypothesis of inadequate strategies seriously,[24] this programme considers police means to be insufficient. In this perspective, the home minister announced his will to amplify the aggressive model of crime control experimented with in Paris since 2002. 10,000 additional CCTV will be installed in the near future and a budget of 30 million Euros will be allocated by the State to develop these technologies in order to detect potential offenders and to anticipate crime. A new task force will also be created to produce knowledge of risk and to 'engage in the war against insecurity in public transportation'.[25]

5 Conclusion: old wine in new bottles or a gradual change?

In *The Culture of Control*, David Garland underlines 'a new and urgent emphasis upon … the containment of danger, the identification and management of any kind of risk' (Garland, 2001, p. 12). In a society structured to manage risk, the police are viewed as knowledge workers placed at the core of a risk communication network. The core technology of such police work is not the use of force, but the gathering and processing of information. The experiences observed in Paris within a department of the 'Police Nationale' confirm the fact that a greater knowledge work dimension is being introduced within police work, either to target the use of force, as in the case of the Paris transport police, or to diversify the spectrum of police tools, as in the example of the Paris urban service. These initiatives are often accompanied by a political rhetoric of 'modernisation' of police activities through the promotion of new methods and uses of information.

However, the statement of newness of these so-called modern police practices has significant limitations. As our French examples suggest (and this observation echoes Manning's observations in the American context), these surveillance-based

23. National Service of Transport Police.
24. As Manning outlines, such saturation patrols that have existed for decades remain dominant police tactics and their efficiency is never questioned even when sophisticated technological means, such as crime mapping, are introduced (Manning, 2008). That is why he concludes that the use of information technology does not lead to significant changes in police practices.
25. Brice Hortefeux, Home Minister, 14 April 2010, http://www.interieur.gouv.fr/sections/a_la_une/toute_l_actualite/securite-interieure/plan-securisation-transports

innovations and the knowledge work dimension may accompany the use of coercive force and legitimate conservative crime control perspectives (see also Purenne & Wuilleumier, 2011). In this French case, the conservative trend illustrated by the rhetoric of 'war against crime' is now firmly established both in public discourse and in concrete measures. Since 2002, a large set of law and order policies has been adopted and certain scholars even raise the idea of a 'frenzy of security measures' (Mucchielli, 2008).

In this light, the risk-management discourse may be conceived as a rationalised expression of professional stereotypes and traditional ways of thinking (Bouagga, 2010). As David Lyon outlines, 'high-tech policing may involve wireless web-based data searches, but it has not moved beyond notion such as city "hotspots" to pinpoint areas requiring special police attention, where "annoying behavior" may occur, or where it is important that certain social elements, above all the poor, be cleaned away for tourism' (Lyon, 2003, p. 21). From that perspective, 'the argument claiming that the police are now an information-based organization, focusing on managing risk and enhancing security, is both premature and flawed' (Manning, 2008, p. 262). Indeed, police commissioners are generally more interested in the symbolic 'aura' of technology, rather than in the potential of surveillance as a means to explore new ways of acting on crime (see also Dupont, 2004).

Nevertheless, even if the shift in police practices is only partial, the far-reaching effects of surveillance-based innovations should not be underestimated. Indeed, even if political leaders or police commissioners want to make an impact in the short term and use surveillance devices to promote a professional image of the police, new ways of thinking can still permeate police forces gradually and can become dominant in the long term. That is why it is important to study the possible changes at stake, both for police institutions and for the 'surveilled',[26] from a more longitudinal or long-term perspective.

References

Beck, U. (1992). *Risk Society. Towards a New Modernity*. London: Sage.

Berlière, J.-M. (1996). *Le monde des polices en France*. Paris: Editions Complexe.

Bouagga, Y. (2010). Qui sont les pires des pires ? Des usages des classifications en Supermax aux Etats-Unis. *Déviance et société*, *34*(2), 201-216.

Didier, E. (2011). Compstat à Paris : initiative et mise en responsabilité policière. *Champ pénal*, 8

26. Many scholars confirm that the consequences of such actuarial logics are not neutral for the surveilled. In particular, the culture of risk management generally results in police over-attention to groups identified as high profile and may have damaging effects in terms of negative stigmatisation of the surveilled (Harcourt, 2007).

Dupont B. (2004). La technicisation du travail policier : ambivalences et contradictions internes. *Criminologie, 37*(1), 107-126.

Ericson, R.V., & Haggerty, K.D. (1997). *Policing the Risk Society*. Oxford: Clarendon Press.

Ferret, J., & Spenlehauer, V. (2009). Does Policing the Risk Society hold the Road Risk? *British Journal of Criminology*, 49 (2), 150-164.

Foucault, M. (1977). *Discipline and Punish*. New York: Pantheon.

Garland, D. (1996). The Limits of the Sovereign State. Strategies of Crime Control in Contemporary Society. *British Journal of Criminology, 36*(4), 445-471.

Garland, D. (2001). *The Culture of Control. Crime and Social Order in Contemporary Society*. Oxford: Oxford University Press.

Hamelin, F., Aust, J., Purenne, A., & Spenlehauer, V. (2007). *La police des chemins de fer. Leçons d'une analyse comparée France/Angleterre*. Rapport de recherche pour l'Institut National des Hautes Etudes de Sécurité.

Harcourt, B. (2007). *Against Prediction. Profiling, Policing and Punishing in an Actuarial Age*. Chicago: University of Chicago Press.

Hood, C. (1983). *The Tools of Government*. London: Macmillan.

Latour, B. (1987). *Science in Action*. Cambridge (Mass.): Harvard University Press.

Lyon, D. (2007). *Surveillance Studies. An Overview*. Cambridge: Polity Press.

Lyon, D. (2003). Surveillance as Social Sorting. Computer Code and Mobile Bodies. In D. Lyon. (Ed.). *Surveillance as Social Sorting. Privacy, Risk and Digital Discrimination*, 13-30. London and New York: Routledge.

Manning, P. (2008). *The Technology of Policing. Crime Mapping, Information Technology, and the Rationality of Crime Control*. New York: New York University Press.

Marx, G.T. (1988). *Undercover: Police Surveillance in America*. Berkeley: University of California Press.

Monjardet, D. (2006). Comment apprécier une politique policière? Le ministère Sarkozy (7 mai 2002-30 mars 2004). *Sociologie du travail*, 48 (2), 188-208.

Mucchielli, L. (Ed.). (2008). *La frénésie sécuritaire. Retour à l'ordre et nouveau contrôle social*. Paris: La Découverte.

Paperman, P. (1992). *Vision en sous-sol. La vie quotidienne des policiers dans le métro. Rapport pour l'IHESI.*

Purenne, A., Aust, J., & Ferret, J. (2006). *La police au bord des routes. La pénétration d'une nouvelle priorité politique au sein des organisations policières. L'exemple de Paris, Lyon, Montréal et Barcelone. Rapport de recherche pour l'ACI 'sécurité routière et société', ministère de la Recherche.*

Purenne, A., & Wuilleumier, A. (2011). L'introduction des technologies de surveillance dans le travail policier. Facteur de changement ou de réassurance ? *Droit et cultures, 61*(1), 9-21.

Renaudie, O. (2008). *La préfecture de police*. Paris : LGDJ.

Roché, S. (2005). *Police de proximité. Nos politiques de sécurité*. Paris: Seuil.

Wood, D. M., & Webster, W. (2009). *Living in Surveillance Societies: The Normalisation of Surveillance in Europe and the Threat of a bad Example*. Paper presented at the European Group of Public Administration Annual Conference, St Julian's, Malta, 2-5 September 2009.

Wuilleumier, A., & Purenne, A. (2011). L'institution policière face aux mises en accusation de la violence d'Etat : vers un renouvellement du répertoire d'action policier ? In D. Laforgue & C. Rostaing (Eds.), *Violences et institutions : réguler, innover ou résister*, 117-136. Paris: CNRS Editions Alpha.

MULTIPLE VIEWS OF DNA SURVEILLANCE: THE SURVEILLED, THE SURVEILLANTS AND THE ACADEMICS[1]

*Helena Machado, Susana Silva and Manuela Cunha**

1 INTRODUCTION

Current state concerns with security and controlling crime have led to an increasing number of governments investing in a variety of bio-information technology to manage the risks posed by criminal elements and terrorist groups. Chief amongst these technologies has been DNA profiling, frequently described as the gold standard for identifying individuals, and thus an important tool in crime prevention, detection and deterrence. The introduction of DNA profiling and databasing into the criminal justice system has assumed the role of a new 'language of truth' (Lynch, Cole, McNally & Jordan 2010). A growing number of countries across the globe are making efforts to develop computerised forensic DNA databases that will enable the authorities to compare profiles from crime scenes and subjects on an automated basis. This kind of 'bio-surveillance' has been satisfying the state's growing appetite for the storage and retrieval of personal information that can be used to identify and classify people.

The management of DNA profiles of 'genetic suspects' (Hindmarsh & Prainsack, 2010) serves the purposes of surveillance and control constructed under the scrutiny of contemporary societies. Different cultural representations and assessments of current and future scenarios created by forensic genetic databases circulate within present-day societies. Debating bio-surveillance is complex as it involves a wide range of ways to capture information about physical and biological traces

1. The research for this chapter was supported by projects coordinated by Helena Machado that are funded by the Foundation for Science and Technology (Portuguese Ministry of Education and Science) through a post-doctoral fellowship (SFRH / BPD / 34143 / 2006), and the project 'Forensic DNA databasing in Portugal: Contemporary issues in ethics, practices and policy' (FCOMP-01-0124-FEDER-009231).

* Helena Machado works at the Department of Sociology, Institute for Social Sciences, University of Minho and Centre for Research in Social Sciences (CICS), Portugal.
Susana Silva works at the Department of Clinical Epidemiology, Predictive Medicine and Public Health, University of Porto Medical School, and Institute of Public Health of the University of Porto (ISPUP), Portugal.
Manuela Cunha works at the Department of Sociology, Institute for Social Sciences, University of Minho and Centre for Research in Anthropology (CRIA-UM), Portugal.

the human body can leave that can ascertain whether a person has been in a particular place or in contact with other people or objects, which might include DNA and fingerprints (trace evidence), iris scans, photographs, or images on CCTV cameras. The most controversial aspects of DNA databases for forensic purposes are related with normative and ethical issues, leading to the question of usefulness versus costs and risks. DNA databases represent a strengthening of the powers of the state for the sake of the collective good – security and tranquillity. In many jurisdictions the establishment and successive expansion of centralised DNA databases for forensic and police uses have been largely based upon efforts and aspirations of key criminal justice and forensic science stakeholders to ensure that more offenders are prosecuted and deterred from re-offending. However, these objectives may imply restricting or limiting citizens' rights, freedoms and privileges. What sort of guarantees can ensure the protection of individual rights and freedoms that are expected in a democratic society – the right to privacy, physical and moral integrity, the right to non self-incrimination and the presumption of innocence?

Different positions reflecting the key ethical, legal, social and cultural implications of DNA databases are shaping patterns of governance and regulation of bio-information at both global and national levels. The stakeholders' arguments reflect both an enthusiasm for its potential benefits in terms of efficient criminal investigation, crime prevention and deterrence, as well as concerns and uncertainties surrounding the social and ethical implications of the uses of genetic information. For instance, some stakeholders, in the areas of politics, forensic sciences and the justice system, maintain that the creation and expansion of DNA databases will help to reduce the need for costly alternatives, such as the use of traditional investigative techniques (McCartney, 2004), and to prevent miscarriages of justice (Garrett, 2010), and that offenders may be deterred from further criminal activity because of the increased risk of detection (Beauregard & Bouchard, 2010). However, academic researchers, advisory bodies, and civil organisations have raised ethical and political considerations that maintain that the primary obligation of the state – the protection of the public from criminal activities – must be exercised with due respect for a number of ethical values, in particular liberty, autonomy, privacy, informed consent and equality (McCartney, Williams & Wilson, 2010).

While scholarly discourses tend to focus on the oppressive elements of DNA profiling and surveillance (see, for example, Cole, 2002; Lazer, 2004; Lynch & McCally, 2009; Lyon, 2007; McCartney, 2004,), political and public discourses usually highlight the benefits of forensic DNA technology in fighting and preventing crime (Hindmarsh & Prainsack, 2010; Williams & Johnson, 2008). In both scenarios, there is little empirical evidence to support their assumptions. For instance, it is yet to be proven that DNA databases for forensic and police purposes have the effect of preventing and deterring crime (Nuffield Council on Bioethics, 2007); and very few published research explores the use of forensic identification evidence and its impacts on the various stages of the criminal justice process (Peterson, Sommers, Baskin & Johnson, 2010).

In this chapter, we will identify the attitudes towards DNA surveillance expressed by a wide range of social actors in the Portuguese context: on the one hand, the perceptions of Portuguese prison inmates; on the other hand, the representations constructed by professionals working for the state in law, forensic genetics and policing, as well as the perspectives of academics working in the fields of law and psychology in this country. More specifically, we aim to embrace complexity and interrelatedness by exploring the many connections constructed by the various actors in mobilising DNA technology for crime prevention and deterrence. We will focus on how ideas are produced through a network of heterogeneous actors (Mopas, 2007) in which social actions are ordered and organised to reproduce the social order and accomplish certain objectives (Manning, 2002). Our analysis will provide insights into the various points of view surrounding the questions raised by the surveilled, the surveillants and academics on issues of inclusion/exclusion, privacy/security and reintegration/suspicion. Our intention is to promote public discussion in order to formulate new guidelines and reconfigure regulation of this 'novel' method of surveillance – DNA profiling and databasing. Thus, we will focus on topics such as assessment of the value of DNA in deterring crime, and the social construction of stigma and the reconfiguration of individual rights and identities which are being forged by the collection, storage and management of genetic information.

2 The legal background

Following international trends regarding the use of genetics for criminal investigation purposes and strategies for the identification, prevention and deterrence of criminal activities, a DNA database was created in Portugal for criminal and civil identification purposes, in accordance with Act 5/2008 of 12 February. The process of inserting DNA profiles into the DNA database has already begun, but it is still very limited in size.[2] Portuguese legislation on the DNA database appears to be restrictive with regard to criteria for the inclusion and removal of profiles, potentially offering better protection for individual rights in comparison to other legal frameworks in Europe (Van Camp & Dierickx, 2007). In fact, Portuguese law only allows for the inclusion of profiles from convicted individuals sentenced to a term of imprisonment of three years or more, unlike other countries (such as, Spain, Sweden, Switzerland and the United Kingdom) where the DNA profiles of mere suspects are inserted into databases.

2. Portugese press news in June 2012 stated that only 600 profiles were included in the DNA database since its creation in 2008. Causes for the lower than expected growth were associated with the restrictive nature of the legislation and financial costs.

In addition, this legal instrument enables the profile of convicted criminals to be removed after judicial proceedings have ended or at the end of the maximum statutory limit for criminal proceedings stipulated in the Penal Code, a maximum of ten years after the sentence has been served (Article 26.1 of Act 5/2008). This contrasts with the legislation in some countries (Austria and the United Kingdom) which allows the profiles of convicted individuals to remain in the database indefinitely, and does not order the destruction of biological samples. Under a separate provision, Portuguese legislation seeks to ensure that each and every individual has the opportunity to resume life in society 'free' from any official record that might directly identify them as a previous offender, in accordance with a humanitarian tradition in the Portuguese penal system that dates back to the nineteenth century (Santos & Gomes, 2003).

In this study we wanted to explore how different groups evaluate the two main topics related to DNA databases: the possible deterrence effect on criminal activities and the fact that the storage of personal data – including DNA profiles – in a centralised database involves potential threats to a range of individual rights, namely the presumption of innocence, as it shifts attention towards *potential* offenders and re-offenders and thus reinforces an atmosphere of prospective policing (Williams & Johnson, 2008). A second topic that we considered challenging to assess, from the point of view of various social actors, was the potential for social rehabilitation and reintegration that the Portuguese forensic DNA database can actually offer, considering that Portuguese legislation has stipulated the removal of the profiles after a certain period of being loaded in the database, in order to insure the opportunity for ex-prisoners to be re-integrated into society with a clean trajectory cleared of any official record that could directly identify him or her as a perpetrator of past crime.

3 Methodology

The first step in gathering the various views on DNA surveillance was to understand the narratives of offenders, since we assumed that the possible deterrent effects that a forensic DNA database may, or may not, provide should also be considered from the prisoners' perspective (Duster, 2006; Prainsack, 2010). Hence a qualitative study based on 31 in-depth interviews with male prisoners in the north of Portugal was carried out between May and September 2009 to survey their perceptions of forensic technology, criminal investigation and surveillance (Machado, Cunha, Miranda & Santos, 2010). We interviewed male prisoners only because approximately 95% of all prisoners in Portugal were male at the time we conducted our interviews.[3] We devised a theoretical sample, based on representativeness by diversity and exemplariness (Hamel, Dufour & Fortin, 1993), combined with convenience sampling by considering the individuals that would be more predisposed to participate in this study, according to the information gathered by the administration in each prison.

We obtained a diversified sample in terms of criminal record (type of crime and duration of the sentencing) and socio-demographic characterisation. At the time the interviews took place, the ages of the prisoners ranged between 21 and 54 years. 17 of them were between 25 and 35 years old. Most participants were single (n=22) and Portuguese (n=30). The education level of the interviewees varied between 4 years of schooling (n=6) and a higher education degree (n=1), but 22 had no more than 9 years of schooling. The professional occupation of most interviewees previous to incarceration was in the secondary sector of the economy, namely in civil construction and industry (n=21).

The prisoners interviewed were mostly primary offenders (n=24) and the crimes which led to their incarceration were: homicide (n=11); rape and/or sexual abuse of minors (n=8); theft (n=8); drug trafficking (n=4); driving without a licence (n=2); and fraud (n=1). The length of the sentences varied between under three years to 25 years. We must emphasise that 20 of the 31 interviewed inmates were given a sentence which was between five and 20 years in prison.

The prison inmates were asked about certain legal provisions concerning DNA databasing in Portugal, namely the criteria for profile inclusion and removal, and the potential of the DNA database to prevent and deter crime, and promote social reintegration. The tapes were transcribed verbatim and transcripts of the interviews were checked for transcription accuracy. Participants' oral consent to conduct and record the interviews was obtained after they were told that they could refuse to answer any question and that anonymity was guaranteed. The qualitative data were then systematically coded and aggregated around the abovementioned two core themes that will be explored in this article. The data was analysed by two researchers and, as the analysis evolved, themes were discussed with a third expert. In the case of conceptual difference, it was reflected upon and resolved by further joint debate.

In September 2010 a one-day 'think tank' was organised with the aim of discussing the main findings of the study conducted with prisoners. The main aim of this meeting was to provide a basis for a reflection that could, perhaps, inform the practices of the professionals involved in the establishment and operation of

3. According to official statistics provided by the Portuguese General Board of Prison Services, on 31 December 2009, 613 (5.5%) out of a total of 11,099 individuals serving prison sentences were female (DGSP, 2009).

the Portuguese DNA database.[4] We invited individuals working for the state in law, forensic genetics and policing, and academics in the fields of law and psychology to attend this discussion. The selection was made on the basis of the relevancy of their work to the establishment and operation of the Portuguese DNA database. Eight experts with the following institutional responsibilities participated in the meeting: a legal expert responsible for the establishment and operation of the law for the Portuguese DNA criminal investigation database; a forensic scientist in charge of managing the DNA database; a legal expert from the supervisory body which controls the activities associated with the DNA database; a member of the criminal investigation police; a forensic genetics laboratory technician; and three academics in the fields of law and psychology.

This meeting aimed to address three primary issues pertaining to the benefits, risks, uncertainties and effects of the use of DNA databasing for surveillance purposes and crime fighting: inclusion/exclusion, privacy/security and reintegration/suspicion. We sought to compare various views on DNA surveillance, as we expected that, from the outset, some would reflect a confident enthusiasm for its potential benefits and the increased efficiency it would bring to criminal investigation work, whilst others would emphasise a shared uneasiness and concern for the uses of this information in terms of citizen privacy, integrity and consent (for example, with regard to the collection of biological samples and the insertion or removal of DNA profiles from the database). The qualitative data were systematically coded and aggregated around the abovementioned core themes that will be explored in this article. The data was analysed by two researchers and, as the analysis evolved, themes were discussed with a third expert.

In short, we aimed to compare views on DNA surveillance expressed by academics, professionals working for the state, and prisoners. The meeting was therefore promoted, not so much with a view to simply presenting the findings of a study on prisoners' perceptions of DNA surveillance issues to an outside audience, but in order to extend the study to include the audience's perceptions as part of the research. This participatory design favoured a practical approach which took the situated nature of the impact of these systems fully into account, both in terms of those who regulate them and who are regulated by them.

4. To pursue the aim of disseminating the findings obtained with the prisoners to the wider community of forensic scientists, criminal investigators, law operators, civil organisations and academics, we produced a report which is publicly available on the internet at http://dnadatabase.ces.uc.pt/

4 Views on profile inclusion and removal criteria

A universal database to counter stigmatisation

Many of the prisoners interviewed (n=12) supported the creation of a universal database covering the entire population, as a means of preventing the police from seeking out suspects on the basis of their profile as a previous offender. In general they discussed certain ethical questions raised by the creation of a DNA profile database covering the entire population, but concluded that the benefits outweighed the idea of repression and the threat to individual rights. This idea was in line with the perceptions of academics and professionals, although they did not expect prisoners to support a universal database, due to obvious concerns about the threats to individual rights in terms of physical and moral integrity, and privacy. This support for the establishment of a universal DNA database reflects the initial plan outlined by the Portuguese government in 2005 to create a genetic database for civil and criminal identification purposes containing the profiles of the entire population (Machado & Silva, 2010). However, while the prisoners and academics stressed the importance of the ethical issues related to the creation of a universal database, the professionals involved in criminal investigation work and forensic genetics highlighted financial and technical matters, such as an undesirable rise in costs associated with the creation of a universal database, especially in the context of the financial crisis.

The arguments presented by the surveilled, the surveillants and the academics are supported by studies on the discriminatory potential of DNA databases. On the one hand, the over-representation of certain ethnic groups in police databases has been systematically noted in literature on the subject (see, for example, Cole, 2002; Duster, 2006; Williams & Johnson, 2008), and also in genetic databases used for criminal investigation purposes (Nuffield Council on Bioethics, 2007). On the other hand, several commentators referred to the possibility that expanding the inclusion criteria for DNA databases may help mitigate the databases' discriminatory potential (Connor, 2003; Haddow, 2008; Kaye & Smith, 2004). These comments have fuelled the debate on the need to reconfigure the issue of the threat to individual rights potentially generated by expanding the inclusion criteria. It therefore appears necessary to continue discussing the matter, whilst ensuring that the debate is, as far as possible, open and accessible to all (Machado, Santos & Silva, 2011; Prainsack, 2010).

In the case of prisoners, it is important to note that the views they expressed on stigma and individual rights in relation to the DNA database are a reflection, to some extent, of the forms of collective stigmatisation that have emerged in crime control in Portugal in recent decades. This may have implications for the prisoners' perceptions of the police and the criminal justice system. These forms of stigmatisation have emerged particularly within the context of pro-active styles of law enforcement. They are by definition more prone to selectivity and bias than reactive methods (Smith, 1997; Willbanks, 1987), and have tended to target specific social and ethnic categories. They have thus compounded the effects of social stratification on crime control (Cunha, 2005; Durão, 2011). Mass procedures were most visible in the patterns for controlling drug-related crime, but were by no means limited to this (Cunha, 2002). Police interventions have been increasingly directed towards specific poor urban territories that have become collective targets for surveillance.

Prisoners' perceptions of the potential effect of DNA technology in creating the stigma of delinquency have already been discussed in a previous study conducted in Austria (Prainsack & Kitzberger, 2009), with different results. These may be explained by the different histories of technology and governance in Austria and in Portugal. Whereas the Austrian forensic DNA database was established in 1997, and is one of the largest in Europe and even in the world (Prainsack, 2008), the Portuguese database is still very limited in size and criminal investigation work is still mainly carried out either on an individual case basis or by resorting to informal genetic databases held by the police (Machado & Prainsack, 2012). The results obtained in Austria indicated that prisoners thought that the DNA database not only added to, but deepened the somatic dimension associated with the stigma of being a convicted criminal (Prainsack & Kitzberger, 2009, p. 70). This may result from the actual experience of being included in the database, but may also be due to the fact that Austrian legislation does not allow for the removal of profiles, except in cases where a suspect is acquitted, if the suspect applies for removal. In Portugal the experience of the stigma associated with inclusion in a DNA database had a prospective character, and was based on the prison inmates' previous personal contact with the authorities' practices regarding known criminals, and their knowledge of how ex-convicts experience difficulties with social reintegration (Machado et al., 2010).

5 Non-removal of DNA profiles to protect individual rights

Portuguese law on the forensic DNA database stipulates the mandatory removal of DNA profiles from the database at the end of judicial proceedings or the end of the maximum statutory limit for criminal proceedings stipulated in the Penal Code. In making the framework and lifetime for DNA profiles equivalent to that of the criminal record, and cancelling the effects of both at the end of the same period of time, the law on the DNA profile database reflects the perspective of the legislators and regulators who planned it on the basis of a notion of stigma that they themselves projected onto the database, by associating it with the criminal record. In making this association, as if both types of data shared the same kind of stigmatising potential from the outset, they inadvertently conferred an equally negative connotation on the DNA database. It is as if a DNA profile inherently contains a criminal identification, rather than merely a personal identification or an individual biological identity.

The prisoners expressed quite a different perspective. Firstly, they identified a more 'neutral' connotation insofar as they did not tend to associate the function and purpose of the DNA profiles with those of criminal records, but instead with fingerprinting – or rather a 'super-fingerprint', in other words a kind of gold standard for individual identification (Lynch, 2003). Secondly, this group of convicted offenders reversed the regulators' arguments on stigma and stigmatising processes. As far as they are concerned, the stigma would precede the factual record anyway, as it would already be attached to people, or more precisely to 'classes' of people: in the view of the interviewees the suspects 'are always the same people'. According to the prisoners, the police can always plant DNA at a crime scene (Prainsack & Kitzberger, 2009; Machado & Prainsack, 2012) as concrete evidence of a previous stigma, but DNA has no stigmatising potential in itself. The stigma is produced by the authorities, and in this sense is independent of the biological marker, and comes into play prior to the collection, storage and management of biological identities. Consistent with this logic of seeing DNA identifiers as just one more element used in criminal investigation practices, the prisoners indicated that the removal of profiles could contribute towards ex-convicts being seen as 'eternal suspects'. The prison inmates considered that the kind of criminal investigation work carried out in Portugal was based much more on fingerprints held in police databases and analysis of the *modus operandi* of criminals rather than on the use of DNA profiling – which is actually true for forensic crime scene investigations all over the world (Schroeder & White, 2009). Hence, the prisoners' previous experiences of the criminal justice system had led to a particular redefinition of their rights and status, in which they viewed the removal of their DNA profiles from the forensic database not as a benefit, but rather as a lost opportunity to prove their innocence and to have a 'clean sheet', 'cleared' of any official record that could directly identify them as the perpetrators of past crimes.

Thus, the prisoners interviewed mainly favoured the non-removal of DNA profiles, with 20 arguing that DNA profiles should never be removed. The arguments for the non-removal of profiles were justified by the 'exonerating' potential of the database (n=6), the claim that the removal of profiles was a hindrance to criminal investigation (n=4), and the fact that removal would limit the possibility of identifying re-offenders (n=7). Profile non-removal could, from the prisoners' point of view, facilitate criminal investigation and also prevent the authorities from suspecting the 'wrong person', whilst proving the innocence of many others. This exonerating potential was perceived both in tangible terms (e.g. determining that it could not have been a certain person who committed a certain act) and in more general and abstract terms (its essentially individual nature would act as a guarantee against a presumption of guilt projected a priori onto collective categories of people).

The professionals shared the prisoners' concerns over the uncertainties associated with the sensitivity of evidence in the sense that this could contribute towards misconduct in investigations. However, the professionals rejected the inmates' idea of DNA being planted at crime scenes, stating that these images were merely 'ghosts' and 'myths' created by prisoners to discredit police work. According to the professionals and academics who participated in this study, two separate issues are involved in the debate on removal criteria: on one hand, the discourse surrounding the law and, on the other hand, the narratives on practices associated with the collection, management and interpretation of evidence. Thus, the discussion of removal criteria should be framed within a broader theme – the mechanisms for controlling and monitoring access to, and use of, the forensic DNA database.

6 DNA profiles, criminal records and social reintegration

The law on the DNA profile database stipulates that DNA profiles be removed at the time of the definitive cancellation of the decisions cited in the criminal record (Article 26.1f of Act 5/2008). As mentioned before, the removal of profiles is intended to reinforce the principle of social reintegration. The removal of profiles, after a maximum period of 10 years after the sentence has been served, has sought to safeguard, from the point of view of the intentions of the legislator, the possibility that each and every individual may be reintegrated into society without any official record that might directly identify them as the author of a past crime (Moniz, 2009). One of the topics that emerged during the 'one-day think tank' debate concerned this particular legal provision. The prison inmates were also asked about their opinion on the equivalent storage time for criminal records and DNA profiles in the national forensic DNA database.

The clear distinction between, on the one hand, the characteristics, function and potential of DNA profiles for identifying perpetrators of crime and, on the other hand, those of the criminal record became more pronounced when seven of the 20 prisoners who opposed the removal of profiles indicated that the criminal record should, on the contrary, be erased (in order to improve job opportunities). Although some prisoners expressed fears about potential abusive practices by the authorities – such as planting evidence at crime scenes in order to incriminate individuals – they thought that the implications of storing DNA profiles permanently in databases were 'neutral', or as neutral as those of fingerprints, and different from those associated with criminal records.

The prisoners interviewed did not believe in the DNA database's potential for social reintegration (Machado et al., 2011), and this opinion appears to be combined with a low level of confidence in the role played by the justice system. This was connected to a general lack of belief in the system's ability to provide opportunities and conditions for social reintegration. The prisoners' expectations of social reintegration were instead associated with issues related to being able to find a job after leaving prison and the type of social networks they would be able to find on the outside that would provide them with various kinds of support. In the opinion of the interviewees, social reintegration is 'impossible' or at least 'very difficult' due to the prevailing ex-convict stigma, which is an obstacle to finding a job and makes relationships with other people difficult (n=19). Some prisoners argued that the removal of information from their criminal record was more important in terms of social reintegration than the actual removal of the DNA profile from the forensic database (n=6).

For the surveilled, the surveillants and the academics, making DNA profiles equivalent to criminal records in terms of their purpose and criteria governing their regulation would be unjustified, particularly with regard to the equal subjection of both to the principle of social reintegration which informs the Portuguese penal system.

7 Perceptions of the value of DNA databases in deterring crime

Around the world, the creation of DNA databases for criminal investigation purposes has been presented by political decision-makers and forensic researchers as having the potential to prevent or deter crime. While the professionals and academics who participated in this study shared this perspective, the dominant opinion in the group of prisoners we interviewed was that the effect on crime prevention or deterrence was questionable or at best only minor. Fourteen prisoners claimed that it did not prevent crime, 7 prisoners said they believed that it may prevent crime and 5 prisoners said that it may or may not reduce crime. Certain reasons were given to justify these opinions, namely the prevalence of crimes perpetrated irrationally or on impulse (n= 9) or the fact that the interviewees believed that many criminals did not care about, or even consider, the possibility of being caught (n= 5).

In addition, some prison inmates claimed that the existence of a DNA database might even lead to an increase in 'professional crime', given that it favours premeditation or better organisation of criminal activities (n= 8), and may encourage criminals to take measures to avoid leaving traces at the crime scene (n= 13) and also lead to more violent crimes as a means of clearing traces from crime scenes (n= 1).

The crime deterrent effect of DNA databases has been proclaimed by some jurisdictions to justify the expansion of criteria for inclusion, namely to address the problem of volume crime. It has yet to be proved that the use of forensic DNA databases has a relevant impact on the prevention and deterrence of crime (Nuffield Council on Bioethics, 2007). Some authors have suggested that if DNA databases can have plausible deterrent effects, translated into decreased crime rates, this effect has not been liable to be independently attributed to the creation and expansion of DNA databases (Bhati, 2010; Briody, 2004; Briody & Prenzler, 2005). Other studies even indicate that the lack of evidence for DNA databases' deterrent effects in the context of volume crime might be explained by offenders becoming more diligent in avoiding traces (Beauregard & Bouchard, 2010; Freckelton & Selby, 2002).

In addition, the system for criminal policy in Portugal favours prevention in terms of general reintegration and individual socialisation (Moniz, 2002). Although the subject of crime prevention and deterrence was not selected, or at least not directly, as a topic of analysis in a similar study carried out in Austria (Prainsack & Kitzberger, 2009), it may be argued that the results obtained from both groups – the prison inmates, professionals and academics – indicated that DNA databases are not perceived as having any particular value and usefulness in terms of preventing and deterring crime.

8 Conclusion

This study produced unexpected and important results that should be considered in terms of the way DNA surveillance is implemented in a national and global context. We have argued here that there is a need for an insight into how surveillance works in concrete terms for a wide range of social actors: those who are surveilled (i.e. convicted individuals), the surveillants (i.e. the agents responsible for the regulation and daily management of forensic DNA databases) and the academics involved in criminology and legal studies. This could serve as a relevant perspective for surveillance studies insofar as it would represent a more practical approach, as well as greater sensitivity towards the impact of surveillance systems on the surveillants. It would thus contribute to a broader, deeper and more nuanced debate on the benefits, risks, uncertainties and effects of the use of DNA databasing for surveillance purposes and crime fighting.

We found similarities and also differences among the two groups whose narratives were analysed: the convicted offenders and the group of professionals working for the state in law, forensic genetics and policing on the one hand, and the perspectives of academics working in the fields of law and psychology, on the other hand.

Both groups shared the view that DNA surveillance does not have any particular value and usefulness in terms of preventing and deterring crime. The prison inmates even emphasised that DNA databases could make criminals more 'professional' or else were not a relevant factor in criminals' analysis of the risk of being caught (for a more detailed discussion, see Machado & Prainsack, 2012).

Views on the criteria for the insertion and removal of the DNA profiles in the national forensic DNA database were more complex, and nuanced by the different actors' particular objectives and position within the social hierarchy. Three main topics emerged during the debate: the possibility of establishing a universal forensic DNA database with the aim of avoiding stigmatisation; the non-removal of DNA profiles in order to protect individual rights; and the questioning of social reintegration promised by the fact that Portuguese legislation has established equivalence between the storage time for criminal records and the DNA profiles held in the national forensic DNA database.

The differences between these two groups were mostly forged by their social status, in terms of access to institutional knowledge and power, in which the distance between the social elites and the prison inmates became quite visible (Prainsack, 2010). While the professionals and academics working in the area of law and genetics emphasised the legal constraints which are the basis of the need to remove DNA profiles, as established in Portuguese legislation, and which also prevent expansion of the criteria for the insertion of profiles (i.e. suspects, and all convicted offenders), in addition to financial and institutional constraints, the prison inmates claimed that a more intensive and extensive use of DNA technology could prove more just in a criminal justice system that is tainted and sometimes corrupted (Duster, 2006, p. 294). While the social elites reflected on social reintegration in broad, abstract terms, the convicted offenders referred to the stigma of being in prison as a mark for life.

Empirical data also suggested a new view on DNA surveillance expressed by those surveilled, namely analysing DNA profiles from the perspective of digital fingerprinting rather than criminal records, which was the dominant view of the surveillants and academics. Instead of a defensive and apprehensive attitude towards certain aspects which could eventually become more problematic, such as the non-removal of profiles, the surveilled tended to view them not as increasing and tightening up a system of control in which they would be the main target, but as a shield that would protect them from systematic and a priori suspicion. The prisoners felt that they were subjected to systematic and a priori suspicion by the human actors in the justice system – either because of their status as convicted criminals and the presumption of guilt that this generates, or because of the low social standing of most prisoners.

This chapter argues that it is necessary to develop an analysis of DNA surveillance that is able to capture the complexity and heterogeneity of its various effects on local, socially and culturally determined contexts. It is necessary to consolidate the tools that capture the hybrid and diverse nature of the effects that may be produced in different audiences, both law-abiding citizens and potential or convicted criminals. For instance, DNA's capacity to provide identification in an objective and secure manner appears to be a factor that may be used to strengthen public trust in the justice system and in criminal investigation agencies. But there remains the need for more comparative studies in order to achieve a better understanding of how prison populations, all of those who have been labelled by the justice system as criminals and individuals whose DNA profiles might be included in forensic databases for criminal investigation purposes are reacting to the growing use of DNA technology in criminal investigation. Similarly, the kind of impact these databases may have on crime prevention and deterrence has yet to be fully understood (Machado & Prainsack, 2012).

References

Bhati, A. (2010). *Quantifying the Specific Deterrent Effects of DNA Databases* [Online: Justice Policy Center, The Urban Institute]. Retrieved September 24, 2011, from http://www.urban.org/uploadedpdf/412058_dna_databases.pdf

Beauregard, E. & Bouchard, M. (2010). Cleaning up your act: forensic awareness as a detection avoidance strategy. *Journal of Criminal Justice, 38*(6), 1160–1166.

Briody, M. (2004). DNA databases and effects on volume crime, in *The Effects of DNA Evidence on the Criminal Justice Process*. PhD thesis. School of Criminology and Criminal Justice, Faculty of Arts, Griffiths University.

Briody, M. & Prenzler, T. (2005). D.N.A. Databases and property crime: a false promise? *Australian Journal of Forensic Sciences, 37*(2), 73–86.

Cole, S. A. (2002). *Suspect Identities: A History of Fingerprinting and Criminal Identification*. Harvard: Harvard University Press.

Cunha, M. P. (2002). *Entre o Bairro e a Prisão: Tráfico e Trajectos*. Lisbon: Fim de Século.

Cunha, M. P. (2005). From Neighborhood to Prison: Women and the War on Drugs in Portugal. In *Global Lockdown: Imprisoning Women, Engendering Resistance*, J. Sudbury (Ed.), (pp. 155-165). New York and London: Routledge,

Connor, S. (2003). Take everyone's fingerprint, says pioneer. *The Independent*, Retrieved March 18, 2010, from http://www.independent.co.uk/news/science/take-everyones-dnafingerprint-says-pioneer-596474.html

DGSP (2009). *Reclusos Existentes em 31 de Dezembro, Segundo a Instrução, Sexo e Nacionalidade* [Existing Inmates on 31 December, According to Education, Sex and Nationality]. Retrieved July 15, 2011, from http://www.dgsp.mj.pt/backoffice/uploads/anuais/20100302020332RecExist_Inst-Sex-Nac.pdf

Durão, S. (2011). Polícia, Segurança e Crime: Ambiguidades e Paixões Recentes. *Etnográfica, 15*(1), 129-152.

Duster, T., (2006). Explaining differential trust of DNA forensic technology: grounded assessment or inexplicable paranoia? *Journal of Law, Medicine & Ethics, 34*(2), 293-300.

Freckelton, I. & Selby, H. (2002). *Expert Evidence: Law, Practice, Procedure and Advocacy*. Pyrmont, N.S.W.: Lawbook Co.

Garrett, B. (2010). The substance of false confessions. *Stanford Law Review, 62*(4), 1051–1119.

Haddow, I. (2008). Debating ethics of DNA database. *BBC News online*. Retrieved March 18, 2011, from http://news.bbc.co.uk/1/hi/uk/7177152.stm (accessed 18 March 2011).

Hamel, J., Dufour, S. & Fortin, D. (1993). *Case Study Methods*. London: Sage Publications.

Hindmarsh, R. & Prainsack, B. (Eds.). (2010). *Genetic suspects: Global governance of forensic DNA profiling and databasing*. Cambridge: Cambridge University Press.

Kaye, D. H. & Smith, M. E. (2004). DNA databases for law enforcement: The coverage question and the case for a population-wide database. In D. Lazer (Ed.), *DNA and the Criminal Justice System: The Technology of Justice* (pp. 247–284). Cambridge, MA: MIT Press

Law 5/2008 (2008). Approves the creation of a DNA profiles database for civil and criminal identification purposes. *Diário da República*, series 1, 30, 12 February 2008, 962-968.

Lazer, D. (2004). *DNA and the Criminal Justice System: The Technology of Justice.* Cambridge, MA: MIT Press.

Lynch, M. (2003). God's signature: DNA profiling, the new gold standard in forensic science. *Endeavour*, 27(2), 93–97.

Lynch, M. et al. (2008). *Truth Machine: the Contentious History of DNA Fingerprinting*. Chicago: University of Chicago Press.

Lynch, M. & McNally, R. (2009). Forensic DNA databases: The co-production of law and surveillance technology. In P. Atkinson, P. Glasner & M. Lock (Eds.), *Handbook of genetics and Society: Mapping the new genomics era*. London: Routledge, pp. 283-301.

Lyon, D. (2007). *Surveillance studies: An overview*. Cambridge: Polity Press.

Machado, H. & Silva, S. (2010). Portuguese forensic DNA database: Political enthusiasm, public trust and probable issues in future practice. In R. Hindmarsh & B. Prainsack (Eds.), *Genetic Suspects: Global Governance of Forensic DNA Profiling and Databasing* (pp. 218-239). Cambridge: Cambridge University Press.

Machado, H., Cunha, M. I., Miranda, D. & Santos, F.. (2010). *Stained bodies. Prisoners' perceptions of the DNA database for criminal investigation purposes and their perspectives of social reintegration*. Retrieved March 18, 2011, from http://dnadatabase.ces.uc.pt/list_documents.php.

Machado, H., Santos, F. & Silva, S. (2011). Prisoners' expectations of the national forensic DNA database: Surveillance and reconfiguration of individual rights. *Forensic Science International, 210*(1), 139-143.

Machado, H., & Prainsack, B. (2012). *Tracing Technologies. Prisoners' Views in the Era of CSI.* Aldershot: Ashgate.

Manning, N. (2002). Actor Networks, Policy Networks and Personality Disorder. *Sociology of Health & Illness, 24*(5), 644–666.

McCartney, C. (2004). Forensic DNA Sampling and the England and Wales National DNA Database: A sceptical approach. *Critical Criminology*, 12, 157-178.

McCartney, C., Williams, R., & Wilson, T. (2010). *The Future of Forensic Bioinformation.* Retrieved May 14, 2012 from http://www.law.leeds.ac.uk/assets/files/research/ccjs/forensic-bioinformation-report.pdf

Moniz, H. (2002). Os problemas jurídico-penais da criação de uma base de dados genéticos para fins criminais. *Revista Portuguesa de Ciência Criminal*, 12 (2), 237- 64.

Moniz, H. 2009. A base de dados de perfis de ADN para fins de identificação civil e criminal e a cooperação transfronteiras em matéria de transferência de perfis de AND, *Revista do Ministério Público, 120* (Oct/Dec), 145–56.

Mopas, M. (2007). Examining the 'CSI effect' through an ANT lens. *Crime, Media and Culture, 3*(1), 110-117.

Moreira, J. J. S. (1994). *Vidas Encarceradas: Estudo Sociológico de uma Prisão Masculina.* Lisboa: Cadernos do Centros de Estudos Judiciários.

Nuffield Council on Bioethics (2007). *The Forensic Use of Bioinformation: Ethical issues*. Retrieved March 18, 2011, from http://www.nuffieldbioethics.org/fileLibrary/pdf/The_forensic_use_of_bioinformation_-_ethical_issues.pdf.

Peterson, J. et al. (2010). *The Role and Impact of Forensic Evidence in the Criminal Justice Process*. National Institute of Justice. Retrieved June 25, 2011, from http://www.ncjrs.gov/pdffiles1/nij/grants/231977.pdf

Prainsack, B. (2008). An Austrian perspective. *BioSocieties* 3(1), 92- 97

Prainsack, B. (2010). Partners in crime: The use of forensic DNA technologies in Austria, In R. Hindmarsh & B. Prainsack (Eds.), *Genetic Suspects: Global Governance of Forensic DNA Profiling and Databasing* (pp. 153–74). Cambridge: Cambridge University Press.

Prainsack, B. & Kitzberger, M. (2009). DNA behind bars: Other ways of knowing forensic DNA technologies. *Social Studies of Science*, *39*(1), 51-79.

Santos B. S. S. & Gomes, C. (2003), *A Reinserção Social dos Reclusos. Um Contributo para o Debate sobre a Reforma do Sistema Prisional*, OPJP/CES: Coimbra, Retrieved March 18, 2011, from http://opj.ces.uc.pt/portugues/relatorios/relatorio_14.html.

Smith, D. (1997). Ethnic Origins, Crime and Criminal Justice. In M. Maguire, R. Morgan & R. Reiner (eds.), *The Oxford Handbook of Criminology*, Oxford: Clarendon Press, pp. 703-759.

Schroeder, D. A. & White, M. D. (2009). Exploring the use of DNA evidence in homicide investigations: Implications for detective work and case clearance. *Police Quarterly*, 12(3), 319-342.

Van Camp, N. & Dierickx, K. (2007). *National Forensic DNA databases – Socio-ethical Challenges and Current Practices in the EU*, Genebanc, European Ethical-Legal Papers, 9, Retrieved September 18, 2011 from https://www.kuleuven.be/cbmer/viewpic.php?LAN=N&TABLE=DOCS&ID=552&PAGE=1

Willbanks, W. (1987). *The Myth of a Racist Criminal Justice System*. Monterey, California: Brooks/Cole.

Williams, R. & Johnson, P. (2008). *Genetic policing: The use of DNA in criminal investigations*. Cullompton: Willan Publishing.

The Boss as Big Brother: Moral Aspects of Workplace Surveillance

*Susanne Wigorts Yngvesson**

'Only one day left of this week in a madhouse.'

This statement has cost a temporary employee at Volvo his job, as he was dismissed after venting his frustration in a Facebook posting. And he was not alone. Two other employees were dismissed after posting comments on Facebook which their employer claimed were prejudicial. Comments of this kind are not new: it goes without saying that people have evaluated their employers throughout history. What makes Facebook postings different, Volvo representatives argued, is 'the number and the kind of people involved in the criticism and comments'.[1] As this example shows, the workplace has expanded as a physical space. The changes brought on by social media mean that communication now takes place in a far larger setting than the coffee room, and employers have to guard their brand carefully. While employees have always been monitored in the workplace, new surveillance technologies are radically more effective, raising the question of their impact on people's working lives and beyond.

1 Purpose, limitations and modes of surveillance

The preceding anecdote exemplifies some of the moral dilemmas relating to workplace surveillance. In this article, I will highlight several more in order to examine their moral and ethical complexities. I will mainly focus on closed circuit camera monitoring, but also refer to other surveillance technologies. Even though I mainly use cameras as examples, I do not suggest that this particular form is any more or less problematic than eavesdropping on employees or reading their

* Susanne Wigorts Yngvesson is Senior Lecturer and Academic Dean at Stockholm School of Theology. She is an expert in issues related to Surveillance Societies, and is doing her main research on ethics of seeing and moral relations, related to CCTV and other aspects of seeing in the surveillance discourse. Her PhD concerned media and professional ethics: *Den moraliska journalisten. En analys av yrkesetik, ideal och dygder* (*The Moral Journalist. An Analysis of professional ethics, ideals and virtues*).

1. Dagens Nyheter: 'Inlägg på Facebook kostade jobbet', http://www.dn.se/nyheter/sverige/inlagg-pa-facebook-kostade-jobbet (2011-02-12).

correspondence. Video monitoring has become a common mode of workplace surveillance, but despite its risks and the issue of privacy violation, its consequences have not been sufficiently analysed. The article is based on literature and Swedish field studies, in which I have sought to apply ethical theories and methods of critical analysis to existing empirical studies in order to lay bare the moral complexities and ethical stakes of workplace surveillance. Although my examples are mainly Swedish, I also address the wider context of international studies of surveillance. Towards the end of the article, I discuss criteria of a moral nature relevant to the evaluation of surveillance. One of my conclusions is that these technologies require not only interpretation of new ethical issues, but a fuller understanding of the agency of the actors involved.

The literature on this subject is dominated by two perspectives as to why people are monitored at work. One is that it protects employee and employer from criminal and other offensive behaviour. The second is that it provides a means of controlling the employee's behaviour and work habits. Having read literature in this field, it seems to me that the reasons invoked largely depend on the kind of work being monitored. There is no general link between the intensity of supervision and the possible threat that the employee may face. For example, camera surveillance is more easily justified in a taxi or a bus than in a newspaper office or church, where security interests are clearly not the primary consideration. From the employee's perspective, monitoring can sometimes be welcome as a form of security. In other cases, it may be perceived as an invasion of privacy. A number of moral issues raise questions about specific methods of surveillance, and when and for what purposes their use is appropriate.

Modes of surveillance operate independently. In practice, individuals can find themselves being monitored by several different kinds of surveillance simultaneously. Today, surveillance studies use a number of terms to describe surveillance technology and its effects. I will not review the various concepts of surveillance here but merely note the complexity of the term. The principal metaphor of surveillance has long been the panopticon, a term derived from the Greek that can be translated as 'all-seeing'. A panopticon is a one-way form of communication that is often intended to limit or prevent injury and violence through a means of controlling, manipulating, and disciplining individuals through involuntary surveillance.[2] Jeremy Bentham developed a panopticon system which he hoped would encourage people to volunteer to learn to be better human beings. In his description, the key elements of the system are, first, a way to see without

2. The distinction was proposed by philosopher Christian Fuchs in a lecture titled 'How should surveillance be defined? Remarks on theoretical foundations of surveillance studies', at the conference A Global Surveillance Society?, April 13–15, 2010. Fuchs differentiates between the English words *surveillance* and *monitoring*.

being seen, and, second, a monitoring system which elicits an intense feeling of being supervised–both elements being as important as physical monitoring.(Bentham, 2002, pp. 72-75) A special system of social surveillance would train criminals to discipline themselves and others, thereby making society a better place for all. The panopticon model was meant to operate in schools, mental hospitals, and so on as a tool of social development. The moral argument in defence of the panopticon was utilitarian: the greatest good for the greatest number. In this view, surveillance could be described as a 'feature of modernity'.(Haggerty & Ericson, 2006, p.4)

In contemporary studies of surveillance, many scholars consider the panopticon to be extremely limited as a concept. Sociologist Kevin Haggerty has called for the walls of the panopticon concept to be torn down in favour of analysing different systems with 'diverse capabilities and purposes'.(Haggerty & Ericson, 2006, pp. 4-5) Other concepts currently in use include the synopticon, whose etymological root comes from the verb 'to watch simultaneously'. A synopticon differs from a panopticon in being a social form of surveillance connected to values such as care, help, and cooperation. An obvious example is Facebook, where surveillance can have unexpectedly synergistic effects. On Facebook, you can be 'friends' with colleagues and managers, who can access each other's posts, whether private or professional, but who are nonetheless expected to follow unstated social and moral guidelines in both spheres. Here is an example: A friend of a friend told me about an employee at a major company in Sweden whose manager had contacted her via Facebook about salary negotiation. She took offence and felt as though she was being monitored, even though she had freely chosen to be 'friends' with her boss on Facebook. The new surveillance technology, and how to use it in the workplace, has developed fast. So fast that legislation and interpretations of human rights issues still have some difficulties in catching up. In the following part I will sketch some relevant aspects of human rights.

2 Freedom from and the right to surveillance

One prominent argument to limit surveillance is based on the notion of rights. Article 12 of the Universal Declaration of Human Rights contains a clause on the right to privacy, as does Article 8 of the European Convention.[3] In the field of rights, there are also related conventions which protect the right of individuals,

3. Article 12: 'No one shall be subjected to arbitrary interference with his privacy, family, home or correspondence, nor to attacks upon his honour and reputation. Everyone has the rights to the protection of the law against such interference or attacks.' Article 8, 1: 'Everyone has the right to respect for his private and family life, his home and his correspondence'.

such as ILO 169, which applies to the protection of employee data. For a rights-based approach, restrictions to the private sphere must be warranted by special reasons. The reason for these restrictions stems from consideration of other rights, primarily those relating to personal security and safety, but also from considerations relating to the rights of freedom of belief and expression.[4] Most debate over the boundaries between interests and rights has centred on the state versus the individual and not on the surveillance of employees by private companies. With the recent development of increasingly sophisticated and accessible surveillance technology, the need for a debate about workplace surveillance has become imperative.

An array of different forms of employee surveillance is today applied on a more or less regular basis: camera surveillance; monitoring of employees' activity in social media; and disciplinary surveillance cultures of time and behaviour. In a study of Swedish attitudes toward privacy in the workplace, fifty-nine percent of respondents felt that workplace monitoring and surveillance had greatly increased in recent years. In addition to revealing that more men than women feel supervised, the survey contained two other findings.[5] First, feeling monitored is not the same thing as actually being monitored, and, second, the sense of being monitored is sometimes welcome. The study indicates a change in Swedish attitudes toward privacy in the workplace, but says nothing about the nature of that change.

While rights-based protection exists for individuals, and employers remain cautious about using potentially intrusive techniques of surveillance, monitoring mechanisms can also be implicit in the technology itself: regulation becomes difficult when surveillance is part of an individual's lifestyle. Termed omnipresent surveillance, this new phenomenon is characterised as a 'calm technology' in which hardware and software become socially invisible. One example is the microchip, now integrated into the fabric of our daily lives and environment. This form of surveillance is already used in some job contexts and it appears to be dissolving the border between private and professional roles, and between technology and the body, because people's access to various systems has become intimately linked with their identity and biometric data.(Lyon, 2009, pp. 17 & 103-106)[6] If

4. Article 3: 'Everyone has the right to life, liberty and security of person'. Article 18: 'Everyone has the right to freedom of thought, conscience and religion; this right include freedom to change his religion or belief/—/' and article 19: 'Everyone has the right to freedom of opinion and expression/—/'. Article 5 in European Convention on Human Rights: 'Everyone has the right to liberty and security of person./—/'
5. TCO granskar: Personlig integritet – även på jobbet, #13/09, s 22.
6. Lyon describes three eras of surveillance technology: the first consisting of a mainframe computer with multiple users; the second, the present, is characterised by personal computers; and the third is an omnipresent or 'calm' technology which will be invisibly integrated into our daily lives.

so, then we will soon enter a third phase of surveillance systems, which will entail major shifts in perspective on issues of integrity, authenticity, and identity. The question will not be about whether one has the right to be monitored or not but about the degree of surveillance that is warranted, who can access information gathered about us, and how such information can be used.

3 Reasons for and against surveillance

There are situations at work when monitoring and surveillance are desirable. For employees, it may be a matter of safety, but perhaps also a matter of registering a particular action or result by making it visible. For employers, surveillance may be desirable in order to protect company data in a flexible labour market in which documents can be copied and distributed in a relatively simple manner. In order to protect information which an employee can access or perhaps even control, employers may require surveillance of both computer traffic and telephone calls. Such monitoring can be perceived as intrusive by employees. There are different interests to take into consideration, which raise two questions. First, where should employers draw the line with surveillance? And second, what data should be considered privacy-sensitive for individuals?

Moral debates about boundaries always imply assessments of consistency and proportionality. Personal integrity, privacy, and surveillance are considered neither wholly justifiable nor wholly improper when compared with other rights such as freedom from torture. Surveillance is required to be carried out 'properly', 'for legitimate purposes', and 'only where necessary'.[7] It would therefore seem to fall to employers to justify the legitimate use of surveillance and monitoring. Arguments for surveillance are justified as exceptions from the norm of the individual's inviolable right to privacy. Yet employers may have reasonable motives for monitoring, such as protecting their property or preventing criminality.

In a study titled *Privacy in the Workplace* produced by the Swedish government, a distinction is made between a situation in which an employer monitors employees who 'are not expecting' to be monitored and a situation in which employees are 'fully aware of being monitored and the scope of that monitoring.' According to the European Court, it is a greater infringement to be monitored surreptitiously.[8] From a moral perspective, our sense of privacy can reasonably be said to have been reduced when we have knowledge of what is happening, and why, as opposed to when we do not have such knowledge. It may create a perception of involvement in the need for surveillance. Yet knowledge of these con-

7. TCO granskar: Personlig integritet – även på jobbet, #13/09, p. 5.
8. *Integritetsskydd i arbetslivet*, p 199.

ditions can also generate an additional, social form of surveillance beyond the technological one. For those being monitored, it can diminish their sense of freedom and autonomy. For employers, it can mean that employees merely exhibit desirable behaviour in situations when they know they are monitored, whereas their doing so in ignorance of being watched would indicate a genuine change in moral consciousness. This tendency in moral behaviour is not related to surveillance in particular. It has been shown in studies of professional research such as journalism, where behaviour-governed ethics in professions with a strongly social collective culture, centres on obedience rather than moral awareness. (Yngvesson, 2006)

If the purpose of monitoring is to prevent crimes or misconduct, the question is: what is gained by monitoring employees? Where employers need to protect certain information, there may be an argument for surveillance, yet the end result may be diminished trust between employers and employees. By this logic, it becomes employers' responsibility to uncover irregularities, rather than employees' responsibility to behave honestly towards their employer. A potential further consequence of the use of surveillance for monitoring effectiveness or misconduct is what Bigo calls the banoptikon-effect, that is to say, a priori suspicion and implicit accusation of those being monitored.(Lyon, 2009, p.54) A company that monitors individuals whose profile indicates a potential liability is effectively disciplining individuals, with the result that everyone feels under suspicion. If the aim instead is to create employee security, this kind of surveillance should improve mutual trust in the form of protection for the employee, rather than safeguards for the employer.

The purpose of surveillance and its methods are two issues that need to be included in evaluations of surveillance. Another way to illuminate the borderzones of surveillance with regard to violations of integrity is to ask whether feelings of personal integrity cannot feasibly be protected from intrusion. Making such an assessment faces a number of difficulties because private sphere and feelings of integrity are diffuse concepts. While there is no clear definition of integrity, it is usually related to values such as autonomy, authenticity, freedom, and self-esteem. Scholars such as Elin Palm have suggested a link between privacy and personal autonomy, an assumption which presupposes that every individual has an inner core of self that is authentic and unique.(Palm, 2008, p. 203) At a normative level, this requires that individuals be consistent with respect to their autonomy and authenticity, and that they assume a responsibility to protect and develop this. I subscribe to a definition of personal space and privacy in terms of autonomy and authenticity, an approach that can also be justifiably motivated by a rights-based perspective. From the autonomy and authenticity perspective, it can be assumed that resistance to unwelcome intrusion into the private sphere necessitates activity of some kind so that the individual can succeed in maintaining control and self-respect. In this case, surveillance and integrity pose not only questions of law–limiting employers or those in power from intruding into the private sphere–but also an existential question about the identity of the individual. Philosopher Charles Taylor describes the modern understanding of authenticity thus:

> 'Being true to myself means being true to my own originality, and that is something only I can articulate and discover. In articulating it, I am also defining myself. I am realising a potentiality that is properly my own. This is the background ... what gives sense to the idea of "doing your own thing" or "finding your own fulfilment"'. (Taylor, 1991, p. 27).

4 Borders and boundaries of integrity

Challenges to the sphere of integrity remind individuals of an authenticity and autonomy that are normally dormant. It is not the employer's surveillance per se but, rather, threats or challenges to the private sphere that are experienced as intrusive. There is reason to distinguish between different kinds of data that, if disseminated, can be more or less sensitive for the individuals concerned. The boundaries may vary between and within individuals, but a structuring of the issue is nevertheless possible. Palm proposes the following schema.(Palm, 2008, 205) First, a challenge to the sphere of integrity relates to clearly sensitive and potentially stigmatising data, such as data related to sickness or sexual orientation. Second, other data are conventionally regarded as being sensitive on a private level but without necessarily being confidential or socially stigmatising; the employee can withdraw from direct or mediated observation or use specific areas as if they were private. An example of that would be a divorce or infidelity. Third, common data that is either sensitive or regarded as private, such as personal code number or someone's salary. The most obvious challenge stems from the first category, with matters such as one's health or family life. At this point, there is often consensus as to what constitutes sensitive data. It is more difficult to judge what belongs to the private sphere in the second category. Palm points out that it is a difference between what an employee knowingly exposes and what she willingly exposes. In the relationship between employer and employee, the employees are more often at a disadvantage regarding protection of private interests, and may accordingly be entitled to invoke the protection of the private sphere.(Palm, 2008, 206)

Discussion of private and public spheres is made difficult by the idea that 'spheres' have an abstract existence, like adjacent rooms that one can enter or leave. In fact, they are a matter of human relationships: how we perceive ourselves and how we are perceived by other people or machines. They have fluid boundaries which vary according to the circumstances. The architects of surveillance systems create categories as a way of addressing the issues raised by the social and political context of surveillance. David Lyon calls it 'social sorting': a way of monitoring and categorising individuals and groups to be included or excluded. Yet it is not clear that these categories are consistent with how we see ourselves.(Lyon, 2009, p.41) On a meta-level, it is worth considering the consequences of the inner logic of surveillance since the technology used to categorise criminals, terrorists, citizens, customers, or employees is the same. Only the purpose of the surveillance differs. However, in what has been dubbed 'liquid modernity'[9] the limits of who we are are also fluid. Surveillance categories serve to identify us, rather than allowing us to identify ourselves.

5 Work as a separate sphere

For many employees, the boundaries between work time and private time, and between the workplace and elsewhere, have become increasingly blurred. Many jobs are now far more mobile than they were just a few decades ago, as regards both physical movement and communication. Work can now take place outside the office, and employers, having provided their staff with laptop computers and phones, expect them to be available at all times. Professions which were always physically mobile, such as drivers of commercial vehicles, have recently experienced an increase in surveillance, for example, CCTV cameras in taxis, buses, subways, and GPS technology for trucks in long-distance traffic. According to David Lyon, there is a paradoxical logic that increased mobility requires new techniques for the 'identification of citizens, consumers and workers within and across national borders'.[10] The feeling of freedom created by an increased mobility in space and time is thus accompanied by increased demands on the part of employers to identify and control those involved in the physical and social areas of the workplace–frontiers that have expanded to become global.

With the blurring of the boundary between professional and personal, it has become difficult to tell where private space begins. Recently, a reporter asked the chief legal officer of a Swedish trade union to define what people are permitted to post on Facebook. The sample statements were divided into three categories: unwise, risky, and foolhardy. Unwise statements include: 'I am so hung over that I can hardly work'. 'X is associated with Y' is a risky statement. Writing

9. The liquid modernity is a concept devised by Zygmunt Bauman: 'The liquidising powers have moved from the "system" to "society". From "politics" to "life-policies"' (Bauman, 2006, p. 11).
10. Lyon, 2009, p. 7.

'Our product is lousy' or 'My boss is a psychopath' is foolhardy. The key factor is whether the statements can be seen as disloyal towards the employer.[11]

When breaches of privacy have been brought to the European Court, the outcome has not drawn a clear distinction between private and professional life. In all the cases in question, the employer has been the state, leaving it unclear whether the verdicts would have been the same if the employer had been from the private sector.[12] Interestingly, however, even when interference in an individual's private life is deemed to be legal there exists no clear distinction between private and professional life, as the authors of *Privacy in the Workplace* note.[13] Even though the area is as yet insufficiently regulated, employers have obvious interests to protect: their motivation to regulate workers' right to privacy clearly is weak since such regulation would likely infringe on loyalty clauses in employment contracts.

The issue is complex because working practices have changed so profoundly; many activities are now closely linked to individual networks and personal information. As a result, an employer can be justified in 'handling personal information about employees'. On a moral and psychological level, however, it remains difficult to generalise about violations of personal integrity. What for one person is an intrusion can be interpreted by another as participation in a community, a private or social sphere, or a professional network. It follows from this argument that there are differing boundaries, depending on the cultural or social context, to what can be considered a private sphere. People occupy different spheres of privacy at work and in their private lives. Even if they are unconcerned about having a protected sphere in their private life, they will in certain work contexts. The private spheres of work thus need to be discussed separately.(Palm, 2008, p.203)

A first step towards delineating the meaning of work and the different motives for surveillance is to separate work from personal life. However, for employers no simple distinction can be made between working hours and private time. We can be private at work, and we can work while we are at home with our family. Employers are not permitted to control or monitor what employees do outside of working hours, unless they are acting in such a way that obliges the employer to take action. Another problematic area in this respect is the physical space of schools, which students are legally obliged to attend but which are a workplace

11. Dagens Nyheter: 'Skitsnack på jobbet kan få sig avskedad', 2011-01-04.
12. *Integritetsskydd i arbetslivet*, pp. 196-199.
13. *Integritetsskydd i arbetslivet*, pp. 207-208. The report argues that rules to protect privacy should be both 'generally applicable for the whole working life' and 'take into account the conflicting interests which arise'.

for their staff. In Sweden, the use of CCTV surveillance in schools is widespread.[14]

There are two main arguments for CCTV surveillance in schools: to protect against violence, theft, and vandalism; and to instil discipline and a feeling of safety among students in the long term. Furthermore, certain commentators have recommended caution as regards what the law deems worthy of protection, apart from potential targets of vandalism or theft. Opponents of CCTV surveillance argue that no one really knows the effects of surveillance in reducing anti-social or criminal behaviour. There is no clear scientific evidence that CCTV has any overall power to reduce crime or vandalism. It is also reasonable to assume that installing CCTV at any particular location merely moves crime elsewhere, where it remains unmonitored.[15] A third argument against CCTV in schools is that it is a substitute for human monitoring. It merely measures crime statistics, rather than supporting preventative measures.[16]

In a court case in 2011, different legal and ethical arguments were ranked in relation to the issue of which spaces warranted supervision. A distinction was drawn between students' lockers and corridors, classrooms, library, and lounges, with the latter two spaces being considered especially deserving of protection for the purposes of students' privacy. With reference to the Convention on the Rights of the Child's paragraph relating to freedom of education, school libraries were designated a CCTV-free area.[17] In spaces such as schools, it was argued that 'The assessment of proportionality should be even stricter in places the public has no

14. In 2007, 332 schools in Sweden carried out monitoring. Some had cameras indoors and outdoors, but most were limited to indoor monitoring. In 2007, Tensta High School had the most cameras, 60 in total.

15. This argument was made by Stockholm's municipal Board of Education when appealing a decision by the Swedish Data Inspection Board regarding CCTV at Bromma High School: 'It is not possible for a school to restrict the areas being monitored since problems will move to another part of the school. Students at Bromma High School stay mainly in the unsupervised areas in the breaks.' Länsrätten i Stockholms län (County Court in Stockholm), Mål nr 21739-08, 2009-09-22.

16. Dagens Nyheter, 2007-11-17; Dagens Nyheter, 2007-12-10; Svenska Dagbladet, 2008-02-05.

17. 'In order that CCTV should not be offensive according to the Personal Data Act it must be clear that there is a clear need for such monitoring and that alternative measures have been tried but found inadequate.' The argument was supplemented by an assessment of interests in which the need for surveillance should 'take precedence over the private interests of the individual'. Kammarrätten i Stockholm (Appeal in Stockholm), Mål nr 6807-09, 2011-01-25. Second argument: Convention on the Rights of the Child, article 16: 'No child shall be subjected to arbitrary or unlawful interference with his or her privacy ....'

access to'.[18] From a moral perspective, however, this legal ruling can seem a rather blunt tool. Even if a reasonable assessment of proportionality can be made, it does not necessarily mean that surveillance (or lack thereof) is perceived as morally acceptable, merely that individuals in specific situations must accept the fact that protection of property outweighs the feeling of intrusion. My point is that a moral evaluation of privacy and abuse is not synonymous with a legal evaluation. A moral evaluation of what merits protection is very different from a legal assessment of the matter. Added to the difficulty of weighing proportionality when assessing issues of integrity is the fact that Swedish law lacks a definition of personal integrity or the private sphere.[19] Instead, it seeks to uphold values, such as privacy and autonomy, that are worthy of protection.

6 Moral aspects of surveillance

Moral arguments about surveillance are dominated by concepts of proportionality and consequence. Their simplified reasoning is that monitoring is acceptable if the needs and interests involved are sufficiently strong. One problem with such reasoning is that the effects of surveillance remain unclear. In the end, it will devolve into a struggle between particular interests. Unlike an approach based on consequences, one based on rights regards certain things as worthy of protection regardless of the consequences. Arguing on the basis of rights is difficult, however, because the various rights–privacy, personal safety, and property–are in practice often in conflict with each other. Ultimately, the criteria to determine which is the most important or least harmful action are also questions of consequences.

How persuasive one finds moral arguments about workplace monitoring depends on one's ethical perspective. The arguments surveyed in this article differ depending on whether one assumes rights-based ethics, consequential ethics, or a combination of both. By assessing the arguments for surveillance in relation to various rights, it becomes possible to formulate a stronger approach, based on

18. This argument is stated by the Education Public Board in the City of Stockholm, in an appeal by the the Swedish Data Inspection Board's decision on CCTVs at Bromma High School: 'It is not possible for a school to restrict the areas monitored, as problems then will be relocated to another part of the school. Students at Bromma High School stay mainly in the unguarded areas of the breaks'. Court of Appeal in Stockholm, Case number 6807-09. The Swedish Data Inspection Board also assesses the degree of privacy violation in relation to the subject of the monitoring as well as its location and time. Datainspektionen (The Swedish Data Inspection Board), Case number 742-2008.

19. Finding of the largest study hitherto of personal integrity in Sweden: *The Protection of Personal Integrity* (Skyddet för den personliga integriteten) 2007, p. 49. The legislative history of the Personal Data Act (PDA) echoes this finding; see proposition 2005/06:173 p. 26f.

the individual, with which to protect specific values such as a private sphere related to property rights. A disadvantage of reasoning exclusively on the basis of rights is that values and structures not defined as rights, such as hierarchies and power structures, run the risk of becoming invisible.

Another option is to weigh the advantages and disadvantages of using surveillance. Those forms of surveillance that are regarded as having optimal consequences can then be defended as reasonable tools that benefit the greatest number. One disadvantage with this procedure is the difficulty of assessing long-term consequences. We may be reasonably skilled at judging the immediate consequences, but it is harder to see the implications of third and fourth generation technology. Another disadvantage of a rigorously consequential assessment is the difficulty of obtaining a neutral analysis of the consequences. Those with commercial interests in building surveillance systems wherever people congregate readily deploy arguments about increased safety. How one approaches the problem will likely be shaped by the kind of response the questioner wants. Most people approve of workplace surveillance as a means of reducing thefts, but they do not approve of being treated as potential objects of suspicion by monitoring technology. Where strong commercial interests are dominant, consequential assessments can be a way of exerting power.

A third option would be a judgement based on rights *and* consequences in which rights establish a framework for values that deserve protection but that do not appeal terribly strong to most people. There should be checks and balances in the use of surveillance in the workplace because the values of autonomy and freedom need to be protected actively. Yet the starting point should be that surveillance, not its absence, is justified. In practice, however, different rights will conflict with one another from time to time; in these cases, it is useful to be able to invoke moral arguments by which to assess the gains and losses for both individuals and groups.

Debates over surveillance have long been dominated by the issue of state surveillance of citizens. In particular, legal rulings and legislation have come to revolve around this issue. When the question is posed in relation to working life, matters are somewhat different. In the workplace, we are not only private individuals with civil rights and obligations, but workers and employers with specific roles related to our workplace identity. Profession involve ideals, rules, behaviour, expectations, demands, and loyalty: factors that serve to construct a sphere that is different from those of private and public. Degrees of surveillance in the workplace can vary enormously, with staff in one area being given total freedom, and those in another being monitored almost constantly.

It is more difficult to assess values when the effects of surveillance methods are considered in relation to shared social values such as democracy. For example: if I know my phone is bugged, does it reduce my confidence in the rights to freedom of speech and whistle-blowing? Does living in an environment monitored by CCTV attenuate our sense of freedom? Integrity is not an individual value exclusively but also a societal interest. Ludvig Beckman has argued forcefully for the importance of democratic factors when delineating the boundaries of personal integrity. Beckman considers especially serious any violation that 'risks undermining people's confidence in their own ability to influence society and therefore undermines their status as equal citizens''.[20] Palm makes a similar argument about the importance of maintaining individuals' autonomy in such a way that they can choose their professional behaviour with regard to their own opinions rather than the prospect of external enforcement.(Palm, 2008, p. 211) When assessing the moral conditions relevant to monitoring in the workplace, the possible effects of shared social values should also be considered. Surveillance technology, writes Taylor, are tools for specific values, not an end in themselves:

> 'What we are looking for here is an alternative enframing of technology. Instead of seeing it purely in the context of an enterprise of ever-increasing control ... we have to come to understand it as well in the moral frame of the ethic of practical benevolence. ... But we have to place this benevolence in turn in the framework of a proper understanding of human agency, not in relation to the disembodied ghost of disengaged reason, inhabiting an objectified machine.' (Taylor, 1991, p. 106).

7 Conclusion

Which moral conditions should determine the regulation of surveillance is partly a question of which aspects seem worthy of protection, and partly a question of which surveillance methods seem reasonable with regard to their consequences for values such as freedom or privacy. Issues worthy of protection are identified as having both an individual and a collective value, often related to safety or protection. However, surveillance and problem-formulation policies identify mechanisms that are more or less characteristic of hierarchical orders. This is an exercise of power that reveals itself in the area of health and safety, for example. In affirming other values, such as dignity, autonomy, freedom, and privacy, it prevents the values of surveillance from dominating or being granted priority over other values.

As regards surveillance hardware– cameras, electronic eavesdropping, or reading of employee email–the moral issue is rather simple. Monitoring that people do

20. *Integritetsskydd i arbetslivet*, 2009, p. 203.

not generally view as an invasion of privacy is permissible if it is in the employee's or the employer's interest to protect or control other values, such as guarding against theft or measuring productivity. Opinions of the boundaries of privacy vary between individuals, but it should not be too hard to find a balance in such cases between what is considered worthy of protection from involvement in working life, and other relevant values that also deserve protecting. However, there seems to be a temptation to trust that the technology itself will function as both a deterrent and a protection against crime. Some studies show that installing CCTV in public places reduces crime; others reveal little impact.[21] One risk about being overly optimistic about surveillance techniques is that they replace human relationships and personal responsibility. Employees no longer need to be morally aware and reflective, they merely need to do the 'right thing' with an eye to what can be monitored. Some forms and methods of surveillance may increase distrust between employers and employees, while other methods may lead to increased feelings of safety and approval.

The moral aspects most actively debated today relate to second generation surveillance systems, i.e. those linked for example to personal computers . With the advent of social media, computer chips, and biometric identity cards, various methods of surveillance are increasingly blurring the boundary between private and professional life, between the workplace and any other location.(Lyon, 2009, p.110-115) Surveillance technologies are increasingly integrated into our physical bodies by means of the so-called 'bodily identity documents' that are being adopted as means of identification by the public and private sectors. We are now wearing surveillance systems, carrying the very machines upon which we depend in order to function at work or as private citizens. Surveillance society is becoming a lifestyle. New surveillance technologies may not change our beliefs about which values are worth protecting, but they will alter the way we think about integrity, freedom, and autonomy. When our bodies become the repositories of information and a key resource in a surveillance society, it is time to analyse what makes us specifically human so as to develop what Taylor calls 'a proper understanding of human agency'.

References

Bauman, Z. (2006), *Liquid Modernity*, Cambridge: Polity Press.

Bentham, J. *(2002)Panopticon. En ny princip för inrättningar där personer övervakas*, Falun: Nya Doxa.

21. For example BRÅ, 2007.

BRÅ (Brottsförebyggande Rådet) (2012-02-25): *Kameraövervakning och brottsprevention. En systematisk forskningsgenomgång*. Rapport 2007:29, http://www.bra.se/extra/measurepoint/?module_instance=4&name=Kamera%f6vervakning.pdf&url=/dynamaster/file_archive/080111/12a1e36b5d22dc5a16bf8f4aad7b4c5f/Kamera%255fsve%255finlaga.pdf

Datainspektionen (2008-10-01): "Beslut efter tillsyn enligt personuppgiftslagen (1998:204) – PuL, Dnr 742-2008.

Fuchs C. "How should surveillance be defined? Remarks on theoretical foundations of surveillance studies",: Lecture at the conference A Global Surveillance Society? (April 13–15, 2010).

Haggerty K D; Ericson, R. V. (eds.) (2006): *The New Politics of Surveillance and Visibility*, Toronto, University of Toronto Press Incorporated.

Integritetsskydd i arbetslivet, (SOU 2009:44), Stockholm, Regeringskansliet.

Kammarrätten i Stockholm (2011-01-25): "Kammarrätten tillåter viss kameraövervakning i gymnasieskolor", (2011-02-18)

Kammarrätten i Stockholm (2011-01-25), Case Nr 6807-09

Lyon, D (2009): *Identifying Citizens. ID Cards as Surveillance*, Cambridge , Polity Press.

Länsrätten i Stockholms län (2009-09-22):Case Nr 21739-08

Palm, E (2008): "Privacy Expectations at Work – What is Reasonable and Why?", *Ethic Theory Moral Prac, Nr 12:pp 201*-215, Springer Science and Business Media, http://managerialcommunication4700.com/technology6.pdf

Skyddet för den personliga integriteten (SOU 2007:22): Stockholm, Regeringskansliet

Svenska Dagbladet (2008-02-05): "Fler skolor ska kameraövervakas"

Taylor, C (1991): *The Ethics of Authenticity*, Cambridge, Massachusetts and London, Harvard University Press.

TCO (2009:13): *Personlig integritet – även på jobbet*, http://www.tco.se/FileOrganizer/TCOs%20webbplats/Publikationer/rapporter/TCO-granskar/2009/1309_Personlig_integritet_i_arbetslivet_w.pdf (from 2010-12-04)

Yngvesson, S W (2006): *Den moraliska journalisten. En analys av yrkesetik, ideal och dygder*, Uppsala, Acta Universitatis Upsaliensis.

Dagens Nyheter (2007-11-17): "Skolorna ska helst arbeta förebyggande"

Dagens Nyheter (2007-12-10): "Kameror inne i varannan skola"

Dagens Nyheter (2011-01-04): "Skitsnack på jobbet kan få sig avskedad"

Dagens Nyheter (2011-02-12): "Inlägg på Facebook kostade jobbet", http://www.dn.se/nyheter/sverige/inlagg-pa-facebook-kostade-jobbet

The securitisation of environmental conflicts, a blessing or a curse?

*Gudrun Vande Walle and Lieselot Bisschop**

1 Introduction

The securitisation of environmental issues and the consequences this holds for the surveillants and the surveilled is the central focus of this article. During the last decennia securitisation of social conflicts has grown exponentially, even in less obvious domains. One aspect that has gone through a securitisation process is the environment. Securitisation of social conflicts often goes together with a criminalisation of these conflicts, which also implies an increase in repressive intervention; a multilateralisation of actors that are given responsibility to take care of security, including taking up of responsibilities for security issues by non-security actors and a socially unequal elaboration of security infrastructure. This chapter explores whether the securitisation of environmental conflicts reflects the same characteristics as the securitisation of urban social conflicts. Based on this analysis we discuss whether securitisation is the right way to address environmental harm.

The central topic of this book is 'Securitisation: the impact on the surveillants and the subject of surveillance'. From a criminology point of view, securitisation is usually related to social conflicts that threaten individuals, such as loitering by youngsters, drinking in public, nuisance, begging, etc. Over time, many of these social conflicts were increasingly perceived and dealt with as security conflicts. Securitisation is a social practice that lifts issues above normal political discussion, disconnects them from questions of social justice in favour of the language of effectiveness, thereby trumping civil liberty considerations (Loader, 2002b, p. 137). This securitisation served as a justification for the enlargement of social control to include new security actors and new technology (Zedner, 2009). Only in the aftermath of 9/11 has criminology succeeded in crossing the borders of

* Gudrun Vande Walle is assistant-professor at University College Ghent, Faculty of Business Administration and Public Administration, Research Unit Governing and Policing Security; and Visiting Professor at Rühr University of Bochum, Master in Criminal Justice, Governance and Police Science. gudrun.vandewalle@hogent.be.
Lieselot Bisschop is a PhD Candidate in Criminology at Ghent University and researcher at University College Ghent, Research group Governing and Policing Security, Faculty of Business Administration and Public Administration. lieselot.bisschop@hogent.be

the classic, locally-oriented security issues towards the securitisation of transnational social conflicts. The fear of new terrorist attacks led to a securitisation of international travelling and various other elements in everyday life: food safety, healthcare and the environment have become part of the security discourse. Social scientists have studied the security discourse and argued that the securitisation process is the wrong reaction to social conflicts. Social conflicts that asked for a more care-based approach or for a restricted application of security measures – as is the case for migration – are now dealt with as security issues (Aas, 2007). It is argued that this securitisation trend comes with perverse effects (Aas, Gundhus Oppen, & Lomell Mork, 2009; Daase & Friesendorf, 2002; Loader, 2002a; Zedner, 2009). However, human security, or those 'threats that endanger the live and livelihood of individuals and communities' (Bogardi, 2004), have not been explicitly discussed in criminology. The question, then, is whether the securitisation of phenomena such as food safety, healthcare or environmental care, that threaten the quality of life of every human being, and of whole communities, should equally be seen as a curse. Or does this securitisation equate to the only guarantee for these issues about the collective quality of life? To answer these questions, this article selected the case of environmental security. Not that long ago the environment was an issue of secondary importance. During recent decades, however, environmental changes have become increasingly prominent topics in contemporary society, and have found their way onto national and international policy agendas. Currently, we even witness environmental change becoming a topic on the security agenda. This chapter focuses on the effects of the securitisation process of environmental conflicts on both the surveillants and the surveilled. It discusses how securitisation affects those actors that take the responsibility to regulate and control environmental issues. Moreover, it pays attention to the possible consequences this holds for those that are being surveilled. First, an introductory chapter on the characteristics of security and securitisation is presented. Then this article discusses how (global) environmental change came onto the agenda of international policy-makers and how the policy initiatives transformed into a securitisation discourse. This will allow us to get a grip on the general developments of environmental change as a policy topic, and may give us an indication of its link with security issues. In the last part, we discuss whether and how environmental security has implications on the surveillants of the environment. We wonder who the key players are in environmental issues and what implications the securitization has for environmental surveillance. We pay particular attention to the context of globalisation and to how global environmental change is a challenge for human security.

2 Security and securitisation in a nutshell

Security: a story about war, begging and food protection

Neutrally defined security refers to the 'freedom from the risk of, loss of, or damage to a thing that is important to survival or well-being' (Barnett, Matthew, & O'Brien, 2010, p. 5). That 'thing' can refer to an object such as an income, a person such as an employee, but also a social system such as a nation state or urban life. Generally, we distinguish between traditional security and human security. The former refers to the protection of the nation state from threats, such as interstate wars, nuclear proliferation or revolution, which can impact on the state's integrity (Liotta & Taylor, 2006, p. 38). It is about keeping the bad guys out (Dalby, 2009) and mainly refers to the control by military and police as surveillants. The latter concept of human security was introduced after the end of the Cold War, because it became increasingly obvious that the main threats did not come from across state borders but from within. Traditional security was perceived as something to protect the state, but failed to protect its citizens against actual threats (Liotta & Taylor, 2006, p. 37). Human security therefore changed in scope from the state to the individual, to include transnational issues as well. Security was no longer about traditionally military issues but included the environment, economy, and other social phenomena. Human security refers to the concern for people and their development: 'the protection of the vital core of all human lives in ways that enhance human freedoms and fulfillment'[1]. The UN Development Programme (UNDP) recognises seven key fields in human security which are: economic security or the right to an assured basic income; food security or the right for all people to have physical and economic access to basic food; health security or the right to basic health service; environmental security or the right to a healthy physical environment; personal security or security from physical violence; community security or the right to be member of a community and the right of communities to exist; and political security or protecting people from human rights violation (UNDP, 1994, pp. 24-33). This human security perspective puts the security of individuals and communities before that of nation states and takes into account the equity of human needs, rights and responsibilities. 'Human security is (…) achieved when and where individuals and communities have the options necessary to end, mitigate or adapt to threats to their human, environmental and social rights, have the capacity and freedom to exercise these options and actively participate in pursuing these options.' (Barnett, et al., 2010, p. 19).

In the practice of research and policy-making this human security approach risks the infinite enlargement of the field of security studies towards all elements that

1. United Nations Trust Fund for Human Security – definition of human security: http://ochaonline.un.org/Home/tabid/2097/Default.aspx (consulted August 14 2011).

impact the quality of life, rendering the prioritisation of security issues rather complex. However, broadening the concept of security beyond the military realm also integrates economic, social, environmental and political concerns such as poverty, resource depletion, economic inequity and environmental changes (Matthew, Barnett, McDonald, & O'Brien, 2010). This allows for a focus on marginalised populations and the social vulnerability of the poor, issues that traditionally tend to be ignored.[2] In face of terrorism threats or the war in Afghanistan, a classic meaning of protection of power and sovereignty sometimes resurfaces, but it is no longer reduced to national borders as recent struggles for democratisation in the Middle East (cf. the Arab spring) have illustrated. Nowadays, security refers to a blurred complexity of the prevention of crime and nuisance, human security and to the protection of national sovereignty, a continuum where either human or traditional security paradigms dominate depending on the angle of incidence. It will often be difficult to detect clearly which paradigm dominates. Water scarcity might seem to be a mere human security issue, but when focusing on the political rationality it can be the initial impetus of war between nations. Air pollution may be defined as a phenomenon that also endangers human security and, in case of transboundary harm, it may touch to national sovereignty.

Securitisation: meaning, characteristics and implications

The transformation of social conflicts into security issues as an ongoing process is called the securitisation process. Relating the notion of security to a social conflict is the result of a securitisation process that happens in the minds of policy-makers and in a specific segment of society. Buzan, Weaver & De Wilde (1998) located the securitisation process on a spectrum of three steps: First, the issue is non-politicised because the state does not deal with it. In step two, a public policy arises which means that the issue becomes politicised. In a third step, the issue is part of a securitisation process which implies that it is presented as an existential 'threat requiring emergency measures and justifying actions outside the normal bounds of politics' (Buzan, et al., 1998, pp. 23-24). For the process to be successful, securitisation needs a securitising actor, a referent object and an audience that accepts the securitisation (Buzan, et al., 1998, p. 36).

Although several texts have been published on the topic, researchers in the field of security and securitisation are not really explicit about the characteristics of the securitisation process. Going through the reflections of different disciplines resulted in a list of characteristics that go together with the securitisation procedure of social conflicts. We found an initial impetus in Bowling and Newburn (2006) and their study of mechanisms that are at play in policing security con-

2. See, for example, the research programmes of the *International Council for Science* (ICSU) on global environmental change (DIVERSITAS, IGBP, IHDP, WCRP).

cerns, and added related characteristics that are mentioned by other researchers in this domain (Aas, 2007; Castells, 2000; Van Calster & Schuilenburg, 2011; Van den Herrewegen & Verfaille, 2011; Zedner, 2007). The following provides a concise overview of three characteristics of securitisation: multilateralisation, the influence of the transnational level and criminalisation. These will serve as a guiding frame of analysis in our discussion of the securitisation of environmental issues.

A first characteristic is the *multilateralisation* of security actors. The responsibility for security moves from state actors, such as the military and the police, to a wide diversity of actors, such as local governments, supra-national institutions, NGOs, citizens, companies, etc. This multilateralisation goes hand in hand with the taking up of security responsibilities by actors who previously conducted non-security tasks. Social workers are an example of a profession that originally aims to cure social conflicts without judgment or repression but has been absorbed in security policy and is expected to support security actors. Citizens are other actors who are given a large degree of responsibility in protecting themselves and their possessions. For instance, the projects of neighbourhood watch that were set up in Belgium in the 1990s exposed a far-reaching responsibility of citizens to protect their neighbourhood in collaboration with the police (Degraeuwe, 2010). Taking up responsibilities seems to work better for middle-class people than for people from lower socio-economic strata. Multilateralisation has evolved into a network of security actors who take collective action and exchange information, and it has even resulted in the blurring of professional rules. An example is the intervention of private security in military peacekeeping operations in order to support the public military forces (Easton, Den Boer, Moelker, & Vander Beken, 2010).

A second characteristic is the increasing importance of the *transnational level*: securitisation of social conflicts into conflicts of human security goes together with the broadening of intervention beyond the nation state level. The consideration of developments on local, national, regional as well as global levels is inevitable given that contemporary society manifests itself through globalised trade and economic interdependence, on-going developments in world-wide communication and the international effects of social conflicts (Aas, 2007; Castells, 2000).

A third characteristic that goes together with securitisation is a *criminalisation* of social conflicts which implies 'an enlargement and strengthening of repression and early punitive intervention' (Van Calster & Schuilenburg, 2011, p. 65; Zedner, 2007, p. 265). Reacting to conflict is no longer a choice but an imperative and criminal justice related responses seem to be necessary (Bowling & Newburn, 2006). In last decade, we have seen the securitisation of social conflicts shifting to an earlier stage of the conflict, before a crime or even before any threat occurs. Examples are the criminalisation of nuisance in city centres and the repressive attitude towards Muslims who want to cross borders (Aas, 2007). This criminalisation process and the related punitive reaction reflect an unequal selection of threatening and non-threatening groups of people, i.e. a segregation of society. This far-reaching security policy is often a panic reaction by authorities after a stream of complaints or after an incident that has had a dramatic impact on society.

In sum, the securitisation of social conflicts was intended to be an answer to the call for more security. However, it is doubtful whether this move towards more security actors, a more repressive attitude and earlier intervention has resulted in more security (Gledhill, 2009). The often discriminating effects of securitisation, and the intrusion on peoples' privacy rights, put the justification of securitisation under pressure. The rest of this article focuses on the securitisation of the environment. After a short illustration of the securitisation process, the characteristics of it are discussed, along the lines of the three criteria set out above. The aim is to analyse whether we see the same negative implications in the securitisation of the environment, and to assess whether the securitisation of environmental conflicts has been a blessing or a curse.

3 The environment on the security policy agenda

As we mentioned earlier, traditional nation state security has shifted to human security, to include issues that threaten the wellbeing of individuals. Safeguarding the environment is one of those issues that has been translated from a social theme to a security matter: a threat to humanity. Environmental security issues refer to pollution and environmental degradation, biodiversity loss as well natural resource depletion. Protecting the environment has, however, not become a security issue overnight. In order track securitisation trends, we should be able to notice a shift from a non-politicised debate to a politicised topic and to the third step of considering the conflict as a social threat that requires emergency intervention (Buzan, et al., 1998). In what follows, we discuss a number of milestones for the emergence of ecology as a policy topic. We provide a brief history of how global environmental change arrived on the security policy agenda. This overview is not meant to be exhaustive, but sketches general trends and milestones, on both global (United Nations) and European level. The final part of this section discusses the shift of environmental issues from non-politicised to politicised and securitised topics.

A history of environmental police making

The 1960s book *Silent Spring* (Carson, 1962) can be seen as the start of the modern environmental movement (Kovarik, 2001). Grounded in scientific facts, this book countered the general enthusiasm about the use of chemical pesticides and was the first book to pay attention to environmental conflicts. By the end of the

1960s, the first attempts for international cooperation on environmental issues manifested themselves with the *UNESCO Conference on Man and his Environment*. The 1970s witnessed continued scientific evidence about the imperilled state of the environment. The controversial report *The limits of growth* (Meadows, Meadows, Randers, & Behrens III, 1972)[3] presented challenging scenarios for global sustainability and warned of the potential world crisis and overburdening of the carrying capacity of our planet if developments in world population, agricultural and industrial production, and pollution and consumption of non-renewable natural resources were not countered. At around the same time, a number of environmental disasters and scandals happen, such as Seveso's chemical explosion (Italy) and Love Canal's toxic waste dumping (Niagara Falls, US). Most importantly however, this decade witnessed the onset of the first environmental policy initiatives and environmental NGOs. Greenpeace and the Worldwatch Institute were founded and, at a policy level, the United Nations Environment Programme emerged, following the UN Conference on the Human Environment of 1972 in Stockholm, and the drafting of the 1973 Convention on the International Transport of Endangered Species of Fauna and Flora (CITES).

In the 1980s, different disasters pointed towards the dangers of industrial pollution: the Bhopal Union Carbide toxic disaster (India), the Exxon Valdez oil spill (Alaska, USA) and the Chernobyl nuclear disaster (Ukraine) (Kovarik, 2001). This went hand in hand with the emergence of more environmental NGOs, such as World Resources Institute and the Environmental Investigation Agency. Pollution was put prominently on the agenda in the 1990s, pushed by many NGOs and intergovernmental organisations. Different policy initiatives saw light of day as a reaction to the growing scientific evidence about climate change. The World Commission on Environment and Development (WCED) published its report entitled *Our Common Future* (Brundtland, 1987) as the first to firmly unite environment and development as a policy issue through its definition of sustainable development as 'development that meets the needs of the present without compromising the ability of future generations to meet their own needs' (Brundtland, 1987). In 1987, the Montreal Protocol of the Vienna Convention for the Protection of the Ozone Layer was also drafted to phase out ozone depleting substances (CFC/HCFC).[4] The most renowned United Nations conferences, treaties and

3. Ever since, these same scholars have been publishing revisions of their work based on newly available data which continue to validate the conclusions of their original work (Meadows, Meadows, & Randers, 1992; Meadows, Randers, & Meadows, 2004).
4. Chlorofluorocarbons and Hydrochlorofluorocarbons.

protocols date back to this decade.[5] In the EU, the environment became more prominent on the policy agenda. The European Commission realised that environmental quality objectives could not avoid the problems from being shifted elsewhere (cf. transboundary pollution) and evolved towards an integrated approach to environmental policy-making. The focus shifted from environmental policy as an addition to other policies towards a focus on combining both economic and environmental objectives. The focus changed towards the entire cycle of production instead of only focusing on promoting end-of-pipe technologies through regulation. There was a boom of environmental legislation from 1996 onwards with a focus on a number of highly political issues. The influence of industry was increasingly – but still slightly – met with more participation by NGOs[6] and other experts. Various private sector initiatives also emerged to improve environmental conditions. Examples are Global Compact, Global Reporting Initiative, ISO 14000 series. Although the environment was undoubtedly on the policy agenda in the 1980s and 1990s and we increasingly saw the involvement of other than governmental stakeholders (NGOs and corporations), it was only from 2000 onwards that environmental issues entered the security discourse. At the beginning of the twenty-first century, the Millennium Development Goals (MDG) set environmental sustainability as a precondition to reach any of the other objectives. The importance of the environment was repeated at the 2002 World Summit on Sustainable Development (Rio+10) and in the Johannesburg Plan of Implementation, which reaffirmed Agenda 21 and the MDG. In 2005, the Kyoto protocol also entered into force. Environmental issues, and climate change in particular, were also increasingly seen as a security issue: the 2003 NATO summit in Valencia mentioned desertification as a security issue; the IPCC's Fourth Assessment Report (2007) stated that environmental changes pose threats, challenges, vulnerabilities and risks for international, national and human security and survival. A few months after that, there was a debate in the UN Security Council about cli-

5. The most important one is definitely the Earth Summit – the *United Nations Conference on Environment and Development (UNCED)* – held in Rio de Janeiro, Brazil in June 1992. This conference resulted in five important environmental programmes (Mazijn, 2009-2010). First of all, the *Agenda 21* action plan, which served as a blue print for global, national and local actions on sustainable development. Second, the Rio Declaration, which lists 27 principles as a guideline for future development of international environmental law that have to varying degrees been incorporated into the legal frameworks of the UN and EU. A third result is the *Statement of Principles on Forests*, meant as a binding convention for the management, conservation and sustainable development of forests, but due US opposition there is to date no legally binding UN convention on forests. The fourth result was the *Framework Convention on Climate Change* which aimed to stabilise greenhouse gasses in the atmosphere and led to the 1997 Kyoto conference and protocol (also rejected by the US). The fifth result of the Rio Summit was the *Convention on Biological Diversity*, which aimed to conserve biological diversity and ensure biodiversity resources are used sustainably and benefits shared equally.
6. During the 1990s the Brent Spar incident and consumer boycott attracted media and political attention.

mate change. This trend continued in the following years when the 2009 UN General Assembly Resolution[7] invited the relevant organs of the United Nations to intensify their efforts to address climate change and requested a report on the possible security implications based on the views of the Member States and relevant regional and international organisations. This security discourse was present in the EU as well (Hey, 2005). They drafted a Directive[8] which requires the Member States to treat breaches of EU environmental legislation as criminal offences. These include the illegal shipments of waste; trade in endangered species or in ozone-depleting substances; the significant deterioration of wildlife habitats forming part of the Natura 2000 network of protected sites; and significant damage to the environment caused by unlawful and intentionally, or with at least serious negligence, emitting substances into the air, water or soil; and unlawfully conducting dangerous activities. This period thus witnesses the emergence of environmental issues on the security agenda.

The environment as a security topic

As we have illustrated above, there is a development in policy-making and social concerns about environmental issues. In fact, ever since the 1960s we have witnessed the expanding presence of the environment as a policy topic, resulting in the contemporary debate on global environmental change involving government actors, corporate entities as well as citizen groups. It has only been since the end of 1980s that the environment has truly become a policy issue. Since the beginning of the twenty-first century, global environmental change has also become a security issue. From a traditional security point of view, environmental changes pose threats to the natural resources of nation states because of possible conflicts that might arise. These can be seen it as a national security threat and military resources may be deployed as a reaction. We can also consider global environmental changes to be international security issues because of the need to prevent environmental conflicts over natural resources. Examples are degradation of drinking water, climate-induced reduction of food production, drought and famine, climate-induced increases in storms and floods and climate-induced migration. Natural resource crimes are a revenue loss for developing countries, and when the revenue goes to rebellious groups it causes destabilisation of the state. Natural resources are in fact often a source of conflict. For low-income countries, wealth of natural resources in many cases makes poor people poorer (Bannon & Collier, 2003; Ross, 2003). A solution is then the proactive application of an integrated policy on economic, environmental as well as development issues. A rather traditional security approach was presented in the *Limits of Growth* report (Meadows, et al., 1972), in which the strategic importance of natural resources in

7. http://www.un.org/News/Press/docs/2009/ga10830.doc.htm
8. Directive 20008/99/EC on the protection of the environment.

conflicts was stressed. This discourse is still present today as witnessed by the importance granted to energy resources for states (Leboeuf & Broughton, 2008). In the traditional security model, government has the central responsibility and government security agencies are to keep within permissible boundaries in order not to harm the social contract. There has, however, been an increasing focus on security issues relating to environmental change that moves beyond the mere nation state paradigm. This is linked to the field of development studies in which the environment is seen as essential for human security because it is a common good and not because of state interests (Matthew, et al., 2010). This broadening of the environmental security concept was also mentioned by the *Our Common Future* report of the Brundtland Commission (1987) and the Rio Summit, and its five initiatives that followed, also viewed environment and development as inextricably connected policy themes. This short historical overview reflects the different dimensions of security: environmental harm as a threat to the nation state and that asks for a military reaction; environmental pollution as a human security issue and environmental harm as a crime issue.

4 Environmental securitisation: characteristics and implications

Multilateralisation in environmental security

As illustrated in the above history of environmental policy-making, an increasing number of stakeholders are involved in environmental security. In the traditional security model, government has the central responsibility. There has, however, been an increasing focus on security issues that move beyond the mere nation state paradigm. This is equally true for environmental issues because these common goods go beyond the mere interest of the nation state (Matthew, et al., 2010). The contemporary debate about the governance of global environmental change involves government actors, corporate entities as well as citizen groups. There is attention for governing these transnational environmental crimes, not only from a government perspective, but also corporate actors pay attention to this. Civil society actors also speak out in the discussion about environmental issues. Actions of NGOs as well as investigative journalist documentaries illustrate the topical value of environmental issues in society. In many security matters, states are no longer the only actors but various actors in different interactions are involved (Wood, 2006). The state still has a role to play in setting the environmental (security) policy agenda, but this role is played in close cooperation with other actors: NGOs, civil society, scientists, international organisations and corporations. International actors have broadened the governance options for environmental issues. There are new actors at work on environmental issues at regional, international or intergovernmental level, and nongovernmental initiatives have also emerged. Various international laws and regulations, but also business self-regulation and management schemes and more complex systems of governance, are at work in order to try to level the playing field for all actors involved. We selected some of the most striking features of multilateralisation. The first one is the dominance of scientists in the debate. Second, this section discusses grassroots securitisation.

Scientists and the environment

In traditional security issues such as nuisance, scientists are not often consulted by policy-makers to set up or to evaluate security initiatives. If they are approached, it is to justify or enforce political decisions that have been made (Vande Walle, Wayenberg, Dormaels, & Easton, 2010). In cases of human security, however, which refers to issues such as environmental security and food safety, science dominates the securitisation process. Scientific knowledge laid the foundation for the prevention of harm, defined the limits of harmless and harmful activities, convinced society of the harm it can cause, and of the importance of lessening the feelings of insecurity related to self-produced risks. This reminds us of Beck's idea about the *scientisation* of risks produced in what he calls our *reflexive modernity* in order to control the hazards of economic production (Beck, 1992). Ever since the publication of *Silent Spring*, (Carson, 1962), natural scientists have had an important function in the translation of their findings into public policy about the environment. In particular, they have influenced public awareness about the use of chemicals and pesticides (White, 2008). Later on, parallel with the securitisation of the environment, a more multi-disciplinary scientific discourse emerged. This involved natural scientists, engineers, economists and social scientists, which each had a role in environmental security. Only since the 1990s can we see environmental change becoming a true security issue on the agenda of social scientists instead of only natural scientists. Global environmental change has – until recently – been rather absent from the security agenda, or at least it was limited to the security threats for the nation states. The security research agenda has generally focused on terrorism, human rights, armed conflicts, HIV/AIDS, landmines, crime, etc. Issues of climate change have not really been addressed from a security point of view. That can be explained by the fact that climate change has been framed as a topic of ecology and exact sciences that focus on doing away with uncertainties about climate change. That does not mean, however, that scientists have reached a consensus about the extent, causes and consequences of environmental harm. First, it is difficult to measure environmental harm. Since the 1960s quite a lot of studies have focused on the relationship between environmental changes (e.g. depletion of natural resources) and violent conflicts (Bannon & Collier, 2003). These have found that environmental factors can have an impact in the generation of conflicts, but that it is difficult to assess their relative contribution (Matthew, et al., 2010). Second, environmental security is a platform of struggle for scientific domination, not in the least because of the huge economic impact environmental regulation can have. An example is the ongoing debate on global warming and the causes of climate change.

The role scientists play in this is, however, very dependent on their place in the policy-making process, and on the social construction of environmental conflicts through mass media and through the mitigation of other interests at stake. The interpretation of environmental harm often changes depending on the person defining it and the criteria being used (White, 2011).

Grassroots securitisation

For traditional security issues, citizens are increasingly given the responsibility to help prevent crime and nuisance. In environmental security as well people display an increased environmental awareness related to a more sustainable lifestyle. We could say we have become more aware of the material impact that our industrial lifestyles have on society. Despite our technological knowledge and higher levels of social organisation, the risks of environmental change are still there and they are there on a bigger scale due to increased production and consumption, and demographic developments. 'Global environmental change poses new and in some case unprecedented threats to human security' (Barnett, et al., 2010, p. 4). Perceived environmental decline contributes to concerns about the environment. Although 9/11 shifted the security focus to terrorism, the environmental concern seems to be both long-lasting and widespread, despite some variation over time, context and persons. People feel insecure about environmental change. This recognition of environmental conflicts is also starting to remodel life in contemporary society because we increasingly think beyond the mere economic, technological, political and social motivations for decision-making and commit to including ecological elements as well (Inglehart, 1995). Consumption patterns are starting to reflect this increase in environmentally awareness (e.g. car sharing, organic products, eco-labels, etc.). This is partially a spontaneous reaction from below and partially a duty imposed by governments.

Similar to what we found for traditional security issues, thinking responsibly about environmental issues is unequally spread depending on the socio-economic and political context. In affluent regions, environmental regulations can set benchmarks and limits. In developing nations, people often do not have the luxury of being aware of it, or are forced to leave their sustainable lifestyle for financial reasons. The impact of environmental conflicts, however, easily travels across time and space and is therefore inherently transnational in both cause and effect (White, 2011). These changes may create new opportunities for some regions or individuals in the world, but it undoubtedly makes others – and especially the less affluent – increasingly vulnerable to adverse consequences for health, social stability and livelihood.

Climate change and global warming are omnipresent today and affect everyone either directly or indirectly. People have become aware of it, either through media coverage or through direct experience of the consequences. However, opi-

nions still differ regarding the causes and solutions to the issue. A focus on environmental issues comes with an inherent ambiguity: one person may perceive something as an environmental conflict whereas another may not. Moreover, issues can be seen as environmental conflicts but could well transpose into other areas as well. Take, for example, the 2010 collapse of the toxic waste reservoir at an aluminium plant in Hungary. This can be perceived as an environmental conflict, a matter of corporate wrongdoing as well as a worker-related accident. The mass media has an important role to play but can also be to blame for making certain issues routine. Some even claim the focus on environmental issues in western society has resulted in 'green fatigue',[9] where citizens stop believing in the relevance of their sustainable lifestyle given the enormous ecological problems. We should also be aware of the framing of environmental topics in mass media. The media's choice to (not) discuss particular issues can be influenced by both the need for enter/infotainment and the link of media with corporations either through stockholders or advertisement income.
It is not evident which actors should have a leading role in environmental conflicts. Sometimes scientists are aware of the imminent dangers whereas at other times local communities know more about the environmental risks they are confronted with. The securitisation of environmental change is therefore sometimes split between the objective security dangers and the subjective security concerns.

Transnationalisation

In the complexity of today's society we are faced with challenges that are inherently transnational. Environmental issues are among those challenges. The risks global environmental change poses to human security are diverse: undermining access to natural resources of water, food, productive soil; threatened rights of livelihood; impacts on economic and social means of existence can be a factor in generating conflict (Barnett, et al., 2010, p. 21). To illustrate this, a framework of environmental flows is useful (Spaargaren, Mol, & Buttel, 2006). Climate change is not limited to national boundaries, nor is pollution, or biodiversity loss. That is the reason why international actors created new options to increase control and governance of environmental conflicts (Leboeuf & Broughton, 2008). In this framework of flows, the governance is inherently transnational. Globalisation makes controls more difficult, for example, in monitoring illegal transport of natural resources or waste. Therefore, we should realise that the nation state security discourse is no longer fit to deal with these issues (Leboeuf & Broughton, 2008). There is an eminent interdependence between the local, regional and global policy-making, which could result in conflict between national and international interests. There seems to be a geographical divide between those actors who have

9. Have you got green fatigue? *The Independent*, 20 September 2007; Enkel Harrison Ford lokt volle zalen op Nagoya, de Morgen, 27 February 2012.

been the main protagonists for the securitisation of environmental change in the 'North' and the rest. Developed countries might consider environmental issues to be security issues to be dealt with yesterday rather than tomorrow. Developing countries, however, are more concerned with their socio-economic development and will support environmental initiatives as long as these do not hamper it. In implementing global environmental policy to counter climate change this has already posed difficulties, such as implementing carbon dioxide emission standards that are perceived as limiting economic development. A similar problem is posed by the promotion of bio-fuels, which are a clear threat for the developing countries because they compete with food production. Some have labelled this eco-colonialism (Leboeuf & Broughton, 2008).

It is beyond doubt that both 'North' and 'South' are at imminent risk of environmental change. The mitigation of and adaption to global environmental change is, however, not equally spread across the globe, and the debate risks being dominated by the most affluent. Given the global dimension of environmental conflicts, limiting actions to national policy will not be effective (Aas, 2007). However, the local impact should not be neglected either because local differences do occur.

Criminalisation or a development toward risk assessment?

Traditional security issues have been the subject of criminalisation and a related repressive response. This is not the case for environmental security. Although criminologists have pleaded to increase of punitive intervention, prevention dominates security policy. That prevention is considered to be more important than repression has a diversity of reasons. Conflicts of human security or human insecurity have recently been translated into security risks (Corry, 2010). This turns a hardly manageable social conflict into an apparently manageable one. Insecurity refers to omnipresent long-term harm. Risk, however, refers to a calculated and rational condition in the future. In order to cope with the harmful side-effects of economic expansion without losing legitimacy, Beck explains how the 'social production of wealth is systematically accompanied by the social production of risks' (Beck, 1992, p. 19). Risks are the scientific-based translation of social conflicts into manageable conditions.

Many international treaties and non-governmental initiatives have been developed to deal with global environmental change. More environmental laws and regulations have developed than we can keep track of. About 270 multilateral environmental initiatives have seen light of day since the 1960s, and many of those have been very successful (e.g. Montreal Protocol, CITES convention or the ISO14000 standard). Despite the existence of numerous environmental treaties and agreements, global environmental governance often tends to be dominated by trade issues. Increasingly, however, the debate is not about how to respond to disasters, but about potential threats. Risk refers to a particular prediction, a potential threat. Applied to the environment, this could refer to biodiversity loss, global warming, desertification, ozone depletion, ecosystem damage, damage to human health and the like. These risks result from production and consumption patterns in our society. Risk is, however, a very subjective thing and it is not always about very visible harm. Environmental change easily impacts across borders (cf. transference of risks) and moves through and accumulates over time. This holds a challenge for interpreting risks and deciding upon appropriate countermeasures and precautions. Many affluent countries are wary of the actual and potential environmental hazards of their consumption or production and therefore choose not to have these hazards located in their own backyard. A way to deal with these uncertainties that was chosen by the international community, and was integrated in many regulatory and legal frameworks, is the precautionary principle.[10] This principle is of central importance in environmental issues because it shifts the burden of proof to those who create the new risks to the environment and who will receive benefits from it. It pressures both state and non-state actors to cooperate in dealing with both positive and negative effects of industrial activity (Maguire & Ellis, 2005).

When an issue is criminalised, we can assume that the government will react to breaches of law involved. The reactions can be placed on a continuum from criminal law to administrative law and to self-regulation, with both controlling (sanctioning, whistleblowers) and persuasive approaches (training, codes of conduct) (Grabosky & Gant, 2000; Ponsaers & Hoogenboom, 2004; Van de Bunt & Huisman, 2004). Assuming that both approaches are not mutually exclusive, Braithwaite (Ayres & Braithwaite, 1992) brought them together in the responsive regulatory pyramid, which tries to avoid state intervention and punishment as long as people succeed in resolving social conflicts themselves. Despite the plethora of environmental instruments at national and international level and the involvement of various actors, criminalisation tends to remain limited to theory rather than being implemented in practice. The implementation of many international conventions is very different across the globe and, even within the EU, efforts of member states are sometimes miles apart. Although much environmentally harmful behaviour has been defined as being criminal, the implementation is generally limited to administrative law or self-regulation, and does not result in a reaction by the criminal justice system.

10. This principle dates back to the 1992 Rio Conference and Declaration. The principles states: 'In order to protect the environment, the precautionary approach shall be widely applied by States according to their capabilities. Where there are threats of serious or irreversible damage, lack of full scientific certainty shall not be used as a reason for postponing cost-effective measures to prevent environmental degradation.'

5 Conclusion: the securitisation of human security: a blessing or a curse?

In the first part of this chapter, we explained the mechanisms that go hand in hand with securitisation of traditional security issues: multilateralisation, transnationalisation and criminalisation. The securitisation of these traditional security issues has not been evaluated in a unanimously positive way. First growing criminalisation and an increase in punitive action has led to indignation. Second the deployment of some non-security actors has given people the feeling that they are being abused for security policy. Other critiques are the thoughtless implementation of far-reaching security measures without preliminary scientific research. After careful consideration, we conclude that the securitisation of traditional social conflicts has missed the aim of creating more security. Our historical overview of environmental policy has shown a slow change of environmental issues from a social conflict to a security issue. We wonder if the securitisation of environmental security may be called a successful enterprise or whether it suffers similar problems. Therefore, we analysed its characteristics on three levels: multilateralisation, transnationalisation and criminalisation.

The multilateralisation of the responsibility for environmental conflicts to include state actors, private actors, citizens, NGOs, scientists, etc. has a positive effect on the implementation of environmental policy. Because we all have a hand in environmental harm it is the task of all to prevent further destruction and to improve the environment. As citizens, however, it is hard to gain correct knowledge about what is harmful and what is acceptable as a risk; some companies lobby to influence scientific opinion in favour of the economic market. Traditional security issues have often been taken as a panic reaction after a disaster or before the elections to reassure citizens and gain their votes. The securitisation of environmental conflicts and food safety is a procedure based on science. Scientists of natural sciences, chemistry, engineering, etc. are in a permanent struggle to establish the scientific truth.

The classic security domain has undergone a criminalisation process. Environmental security is more easily translated into risks. The message is that we all have to take care of our planet by limiting our ecological footprint. That makes the idea of prevention and the precautionary principle more prominent than repressive reaction. Companies and citizens all have the chance to work in a positive atmosphere to resolve environmental conflicts. Compared to the more traditional security issues, the excesses of criminalisation have been avoided here. The balance may, however, be inclined too much towards risk calculation and adapted prevention. Criminologists have criticised this attitude of impunity as being insufficiently rigorous in the face of the most drastic violations of human security. Human insecurities such as food safety and environmental crime, which often affect many people, are hardly ever met with a repressive response. Apart from the translation of harm into risk, it may be explained by an ineffective state that has passed on its responsibilities to scientists, companies, NGOs and citizens without keeping its guiding authority. That makes human security vulnerable to fluctuations. The protection of the environment has been made dependent on the willingness of companies and citizens to take precautionary measures. In times of economic crisis, or when people start to suffer from 'green fatigue', there is no authority that takes the reins.

The securitisation of social conflicts has often been criticised by social scientists because it stimulates targeted gentrification, it is a direct attack on civil rights, it means a displacement of specific groups and it is a step back in the democratisation process (Aas, 2007; Bigo & Elspeth, 2005; Wilson & Weber, 2008). The securitisation of human security is in itself more democratic because it improves the wellbeing of all. It is hard to clean the air of the rich without positive effects on the air of the poor, or to clean the seas of the Western countries while leaving the other seas polluted. This holds true up to a certain the level: polluting industries are transferred to third world countries or electronic waste of the western market can be dumped. Although environmental change of course poses threats in itself, it is mainly the interplay of these with social and political processes that cause disasters. Reducing poverty, preventing conflicts and controlling environmental change are prerequisites to sustainable development. Therefore sustainable development and human security policy-making should be intertwined. Traditional understandings of governing security are limited and we should rather focus on an integration of security, development and sustainable development paradigms in both research and policy-making. We should also, as mentioned before, be aware of the strengths of other than state actors in governing environmental issues and should therefore not only focus on causes of insecurity. The 'Northern' or industrialised frame of reference is not a panacea for all either.

Compared to the securitisation of classic social conflicts, the balance of the securitisation of environmental security is more positive. More attention goes to the vulnerability of collective goods, it has become a responsibility for all, the emphasis is on prevention on a transnational level and it is scientifically based. This securitisation procedure may be considered as a constructive example for classic security issues. We have found some weaknesses that are less prominent for classic security issues. The most important is the withdrawal of the state, which makes it a domain that is highly dependent on the taking up of responsibilities by non-state actors. We can conclude by saying that security researchers may have to work closer together to exchange ideas about protecting security.

References

Legal and policy documents (in chronological order)

Convention on International Trade in Endangered Species of Wild Flora and Fauna, with Appendices, Washington, 3 March 1973 (entered into force 1 July 1975)

Montreal Protocol on Substances that Deplete the Ozone Layer, Montreal, 16 September 1987 (entered into force 1 January 1989)

Basel Convention on the Control and Transboundary Movements of Hazardous Waste and their Disposal, 22 March 1989 (entered into force 5 May 1992)

United Nations Framework Convention on Climate Change, New York, 9 May 1992 (entered into force 21 March 1994)

Rio Declaration on Environment and Development, adopted at the UNCED in Rio de Janeiro, Brazil, 3-14 June 1992

Agenda 21, The United Nations Programme of Action from Rio, Earth Summit, adopted at the UNCED in Rio de Janeiro, Brazil, 3-14 June 1992

Statement of principles for the Sustainable Management of Forests (Annex III: Non-legally binding authoritative statement of principles for a global consensus on the management, conservation and sustainable development of all types of forests), adopted at the UNCED in Rio de Janeiro, Brazil, 3-14 June 1992

United Nations Convention on Biological Diversity, adopted at the Nairobi Conference on 22 May 1992 with the, (entered into force on 29 December 1993)

Literature

Aas, K. F. (2007). Analysing a world in motion – Global flows meet 'criminology of the other'. *Theoretical Criminology, 11*(2), 283-303.

Aas, K. F., Gundhus Oppen, H., & Lomell Mork, H. (2009). *Technologies of insecurity.* Oxon: Routledge-Cavendish.

Ayres, I., & Braithwaite, J. (1992). *Responsive regulation: transcending the deregulation debate.* New York: Oxford University Press.

Bannon, I., & Collier, P. (Eds.). (2003). *Natural resources and violent conflict.* Washington DC: International Bank for Reconstruction and Development / The World Bank.

Barnett, J., Matthew, R., & O'Brien, K. (2010). Global environmental change and human security: an introduction. In R. Matthew, J. Barnett, B. McDonald & K. O'Brien (Eds.), *Global environmental change and human security* (pp. 3-32). Cambridge (MA) – London (UK): The Massachusets Institute of Technology Press.

Beck, U. (1992). *Risk society: Towards a new modernity.* London – Thousand Oaks: Sage.

Bigo, D., & Elspeth, G. (2005). Policing in the name of freedom. In D. Bigo & G. Elspeth (Eds.), *Controlling frontiers. Free movement into and within Europe* (pp. 1-9). Aldershot: Ashgate.

Bogardi, J. (2004). Bogardi J., Hazards, risks and vulnerabilities in a changing environment: the unexpected onslaught on human security. *Global Environmental Change, 14*, 361-365.

Bowling, B., & Newburn, T. (2006). *Policing and National Security*. Paper presented at the London-Columbia 'Police, Community and Rule of Law' workshop.

Brundtland, G. H. (Ed.). (1987). *Our common future: The World Commission on Environment and Development*. Oxford: Oxford University Press.

Buzan, B., Weaver, O., & De Wilde, J. (1998). *Security : A New Framework for Analysis*. Boulder: Lynne Rienner.

Carson, R. (1962). *A Silent Spring* (Vol. Harcourt): Houghton Mifflin

Castells, M. (2000). *The Rise of the Network Society, The Information Age: Economy, Society and Culture Vol. I* (2 ed.). Cambridge (MA)/Oxford (UK): Blackwell Publishers.

Corry, O. (2010). *Securitzation and 'Riskization': Two Grammars of Security*. Paper presented at the 7th Pan-European International Relations Conference.

Daase, C., & Friesendorf, C. (2002). Introduction: Security governance and the problem of unintended consequences In C. Daase & C. Friesendorf (Eds.), *Rethinking security governance* (pp. 1-20). London: Routledge.

Dalby, S. (2009). *Security and environmental change*. Cambridge: Polity Press.

Degraeuwe, T. (2010). Samen onveiligheid te lijf. Lokale burgerinitiatieven inzake criminaliteit en onveiligheid. *Panopticon, 6*, 37-53.

Easton, M., Den Boer, M., Moelker, R., & Vander Beken, T. (Eds.). (2010). *Blurring Military and Police Roles. Boom Juridische Uitgevers, Reeks Het groene gras, Den Haag, 2010*. Den Haag: Boom Juridische Uitgevers.

Grabosky, P., & Gant, F. (2000). *Improving environmental performance, preventing environmental crime*. Canberra: Australian Institute of Criminology Research and Public Policy Series.

Hey, C. (2005). EU Environmental Policies: A short history of the policy strategies. In S. Scheuer (Ed.), *EU Environmental Policy Handbook. A Critical Analysis of EU Environmental Legislation* (pp. 17-30). Utrecht: International Books.

Inglehart, R. (1995). Public support for environmental protection: Objective problems and subjective values in 43 societies. *PS: Political Science and Politics, 28*(1), 57-72.

Kovarik, W. (2001). Environmental History Webpage. Retrieved June 20, 2011, from http://www.environmentalhistory.org/

Leboeuf, A., & Broughton, E. (2008). *Securitization of Health and Environmental Issues: Process and Effects. A Research Outline*. Paris/Bruxelles: Institut français des relations internationales.

Liotta, P., & Taylor, O. (2006). Why human security? *The Whitehead Journal of Diplomacy and International Relations, Winter/Spring*, 37-54.

Loader, I. (2002a). Policing, securitization and democratisation in Europe. *Criminology and Criminal Justice, 2*(May), 125-153.

Loader, I. (2002b). Policing, securitization and democratization in Europe. *Criminology and Criminal Justice, 2*, 125-153.

Maguire, S., & Ellis, J. (2005). Redistributing the burden of scientific uncertainty: Implications of the precautionary principle for state and non-state actors. *Global Governance, 11*(4), 505-526.

Matthew, R., Barnett, J., McDonald, B., & O'Brien, K. (Eds.). (2010). *Global environmental change and human security*: Massachusetts Institute of Technology.

Mazijn, B. (2009-2010). Duurzame ontwikkeling: socio-ecologische aspecten in mondiaal perspectief. Universiteit Gent: Vakgroep Studie van de Derde Wereld.

Meadows, D. H., Meadows, D. L., Randers, J., & Behrens III, W. (1972). *the Limits of Growth: A Report for the Club of Rome's project on the Predicament of Mankind*. New York: Universe Publications.

Ponsaers, P., & Hoogenboom, B. (2004). Het moeilijke spel van wortel en stok. Organisatiecriminaliteit en handhaving-strategieën van bijzondere inspectie- en opsporingsdiensten. *Tijdschrift voor Criminologie, 46*(2), 165-181.

Ross, M. (2003). The Natural Resource Curse: How wealth can make you poor. In I. Bannon & P. Collier (Eds.), *Natural resources and violent conflict*. (pp. 17-42). Washington DC: International Bank for Reconstruction and Development / The World Bank.

Spaargaren, G., Mol, A., & Buttel, F. (2006). *Governing environmental flows. Global challenges to social theory*. London/Cambridge (MA): MIT Press.

UNDP. (1994). *Human Development Report*. New York Oxford: Oxford University Press.

Van Calster, P., & Schuilenburg, M. (2011). Emergente processen in het veiligheidscomplex. *Panopticon, 1*, 65-70.

Van de Bunt, H., & Huisman, W. (2004). Organisatiecriminaliteit. *Tijdschrift voor Criminologie, 46*(2), 106-120.

Van den Herrewegen, E., & Verfaille, K. (2011). Waarheidsaanspraken over 'onveiligheid'. De kloof tussen objectieve en subjectieve onveiligheid anders bekeken. *Panopticon, 3*, 4-20.

Vande Walle, G., Wayenberg, E., Dormaels, A., & Easton, M. (2010). De plaats en betekenis van evaluatieonderzoek in de interbestuurlijke veiligheids-en preventieplannen *Panopticon, 3*(31), 40-57.

White, R. (2008). *Crimes against nature: Environmental criminology and ecological justice*. Cullompton: Willan Publishing.

White, R. (2011). *Transnational environmental crime. Towards an eco-global criminology*. London & New York: Routledge.

Wilson, D., & Weber, L. (2008). Surveillance, Risk and Preemption on the Australian border. *Surveillance & Society, 5*(2), 124-141.

Wood, J. (2006). Research and innovation in the field of security. In J. Wood & B. Dupont (Eds.), *Democracy, Society and the Governance of Security* (pp. 217-240). Cambridge/New York: Cambridge University Press.

Zedner, L. (2007). Pre-crime and post-criminology? *Theoretical Criminology, 11*, 261-281.

Zedner, L. (2009). *Security*. Abingdon: Routledge.

About the Author

Gudrun Vande Walle has a master degree in Law (KULeuven) and in Criminology (UGent) and obtained her PhD in Criminology at Ghent University in 2003 on the topic of conflict resolution for victims of corportate crime. She is Assistant Professor at University College Ghent (affiliated Ghent University), Faculty of Business and Public Administration and visiting professor at Rühr University of Bochum, Master in Criminology, Governance and Police Science. She is also member of the research unit Governing and Policing Security (GaPS). Research and teaching experience covers: Corruption and anti-corruption policy in the private and public sector; ethics and integrity; surveillance technologies; conflict resolution and financial-economic Crime. For more details, see: http://ugent.gaps.be/

Evelien Van den Herrewegen has a master degree in Sociology (KULeuven) and in Criminology (UGent) and obtained her PhD in Criminology at Ghent University in 2011 on the topic of people's perceptions of safety. Currently she is involved in a study that explores an integrated approach to security in the port of Antwerp and is lecturer at Courtrai University College in the Bachelor Social Safety. She is also member of the Research Unit Social Analysis of Security (Ghent University). Her research interests are social security questions in late modernity; interpretative research; active citizenship and social cohesion. http://be.linkedin.com/pub/evelien-van-den-herrewegen/4/197/a6b

Nils Zurawski studied Sociology, Ethnology and Geography in Münster, Germany. PhD on "Virtual Ethnicity. Studies on Identity, Culture and Internet" (Peter Lang 2000). Other research experiences are: fieldwork in Northern Ireland on "Violence and Identity" (2000/2001); research project on CCTV and spatial perception (2003-2006), research on consumption and loyalty cards (2008-2010). Research interest include surveillance (CCTV, consumption, theory, anthropology of surveillance), identity, urban studies, space, political anthropology, violence, Internet and media, qualitative methods. He is an MC member of the COST action "Living in Surveillance Societies". He currently works as a senior researcher and professor at the University of Hamburg, More info: http://www.surveillance-studies.org/zurawski/

Het groene gras

Het groene gras ('the Green grass') is a book series that primarily focuses on comparative criminological work, both comparative between the Netherlands and Belgium, as well as outside these countries. The series has a multi-disciplinary set-up, whereby the criminologists are given a prominent place in relation to other disciplines. In addition to Dutch language publications, English language works can also take their place in this series. The book series will be supervised by an editorial board comprising of Stefan Bogaerts, Willy Bruggeman, Paul De Hert, Monica den Boer, Marleen Easton, Henk Elffers, Wim Huisman, Jeroen Maesschalck, Patrick Van Calster, René van Swaaningen, Tom Vander Beken en Paul Ponsaers. The editorial board safeguards the quality of the submitted manuscripts thanks to the peer-review procedure, and develops a proactive policy with a view to the realization of comparative studies. Furthermore, *Het groene gras* wishes to provide doctoral thesis manuscripts that complement the editorial line of the series.

The following parts appeared in this series:

- Lieven Pauwels, *Buurtinvloeden en jeugddelinquentie* (2007)
- Isabel Verwee, Paul Ponsaers & Els Enhus, *'Inbreken is mijn vak'* (2007)
- Freya Vander Laenen, *Drugpreventie bij kwetsbare groepen?* (2008)
- Marleen Easton, Lodewijk Gunther Moor, Bob Hoogenboom, Paul Ponsaers & Bas van Stokkom (eds.), *Reflections on Reassurance Policing in the Low Countries* (2008)
- Lex Cachet, Sofie De Kimpe, Paul Ponsaers & Arthur Ringeling (eds.), *Governance of Security in the Netherlands and Belgium* (2008)
- Stephan Parmentier & Paul Ponsaers (red.), *De Vlaamse advocaat: wie, wat, hoe?* (2008)
- Gudrun Vande Walle & Patrick Van Calster (red.), *De criminologische kant van het ondernemen* (2009)
- Joanna Shapland & Paul Ponsaers (eds.), *The informal economy and connections with organised crime: the impact of national social and economic policies* (2009)
- Lieven Pauwels (ed.), Wim Hardyns & Maarten Van de Velde, *Social disorganisation, offending, fear and victimisation. Findings from Belgian studies on the urban context of crime* (2010)

- Marc Cools, Sofie De Kimpe, Arne Dormaels, Marleen Easton, Els Enhus, Paul Ponsaers, Gudrun Vande Walle &Antoinette Verhage (eds.), *Police, Policing, Policy and the City in Europe* (2010)
- Stefaan Pleysier, *'Angst voor criminaliteit' onderzocht. De brede schermerzone tussen alledaagse realiteit en irrationeel fantoom* (2010)
- Marleen Easton, Monica den Boer, Jelle Janssens, René Moelker & Tom Vander Beken (eds.), *Blurring Military and Police Roles* (2010)
- Inge Vanfraechem, Ivo Aertsen & Jolien Willemsens, *Restorative Justice Realities. Empirical Research in a European Context* (2010)
- Ludo Block, *From Politics to Policing. The Rationality Gap in EU Council Policy-Making* (2011)
- Tom Van den Broeck, *Formalisering/informalisering van sociale controleprocessen. 20 jaar politie- en veiligheidsbeleid in perspectief* (2012)
- Diederik Cops, *Angst voor criminaliteit bij jongeren. Een geïntegreerde benadering van een vergeten sociale groep* (2012)
- Hanne Op de Beeck, *Strain en jeugddelinquentie: een dynamische relatie? Een toets van twee centrale verklaringsmechanismen uit Agnews General Strain Theory* (2012)
- Anne R. Van Ewijk, *Diversity in Uniform. A Conceptual Exploration of Diversity and an Empirical Investigation of Diversity in the Police Context* (2012)
- Gudrun Vande Walle, Nils Zurawski & Evelien Van den Herrewegen, *Crime, Security and Surveillance. Effects for the surveillant and the surveilled* (2012)